Authority, Process and Method

Jewish Law in Context

A series edited by Neil S. Hecht

Volume 1
JUDICIAL DEVIATION IN TALMUDIC LAW
Governed by Men, Not by Rules
Hanina Ben-Menahem

Volume 2
AUTHORITY, PROCESS AND METHOD
Studies in Jewish Law
Edited by Hanina Ben-Menahem and Neil S. Hecht

This volume is a joint project of

The Institute of Jewish Law
Boston University School of Law
765 Commonwealth Avenue
Boston, MA 02215
Publication No. 24

The Institute for Research in Jewish Law
Faculty of Law
The Hebrew University of Jerusalem
Mount Scopus, Jerusalem
Publication No. 25

This book is part of a series. The publisher will accept continuation orders which may be cancelled at any time and which provide for automatic billing and shipping of each title in the series upon publication. Please write for details.

Authority, Process and Method

Studies in Jewish Law

Edited by

Hanina Ben-Menahem
Faculty of Law
The Hebrew University of Jerusalem

and

Neil S. Hecht
Boston University School of Law
Boston, Massachusetts

CRC Press
Taylor & Francis Group
Boca Raton London New York

CRC Press is an imprint of the
Taylor & Francis Group, an **informa** business

First published 1998 by Harwood Academic Publishers

Published 2018 by CRC Press
Taylor & Francis Group
6000 Broken Sound Parkway NW, Suite 300
Boca Raton, FL 33487-2742

ISBN-13: 978-90-5702-552-5 (hbk)

British Library Cataloguing in Publication Data

Authority, process and method : studies in Jewish law. –
(Jewish law in context ; v. 2)
1. Jewish law
I. Ben-Menahem, Hanina II. Hecht, Neil S.
296.1'8

We express our appreciation to Beatrice, Jay and Judith Espovich
through whose generosity this volume has been published
in memory of The Honorable Norman Espovich

Contents

Acknowledgments

The papers in this volume originally appeared in the following volumes of the *Shenaton Ha-Mishpat Ha-Ivri*, the annual of The Institute for Research in Jewish Law of The Hebrew University of Jerusalem: chapter 1, vols. 6–7 (1979–1980), pp. 105–132; chapter 2, vols. 9–10 (1982–1983), pp. 51–71; chapter 3, vols. 9–10, pp. 265–297; chapter 4, vols. 6–7, pp. 405–423; chapter 5, vols. 14–15 (1988–1989), pp. 1–30; chapter 6, vol. 8 (1981), pp. 309–355; and chapter 7, vol. 1 (1974), pp. 72–97.

Chapters 1–6 were translated by Shmuel Wosner; chapter 7 was translated by Tzipora Brody. The translations were edited by Nessa Olshansky-Ashtar, who translated the postscript to chapter 4 and assisted us in the onerous task of preparing this volume for publication.

Introduction

Authority, Process and Method: Studies in Jewish Law, the second volume in the series *Jewish Law in Context*, brings to English-speaking readers a number of seminal articles originally published in Hebrew in *Shenaton Ha-Mishpat Ha-Ivri*, the annual of The Institute for Research in Jewish Law of The Hebrew University of Jerusalem. The authors were given an opportunity to examine the translations and to update or amend their views; these articles therefore represent their latest positions on the subjects in question. The studies we have included have a general jurisprudential dimension yet exemplify salient features of Jewish law. These criteria reflect the goals that motivate the entire series—to present Jewish law in its own context as well as in the context of contemporary jurisprudence. The series' inaugural volume, for example, *Judicial Deviation in Talmudic Law*, explores the theory of adjudication in Jewish law, highlighting its divergence from the positivist model.

The following comments do not constitute a summary of the articles. Rather, they point to thematic connections running through the entire volume, especially those arising from the relations between authority, process and method, and allude to some of the broader questions raised.

Awareness of the distinctive nature of the decision-making process in Jewish law can contribute significantly to our understanding of its philosophy and historical development. The first four chapters analyze aspects of the decision-making process; each of the last three presents an analysis of a general jurisprudential idea as it is conceived in Jewish law. The first article, by Warhaftig, is unique in that in addition to this analysis, it also provides a fascinating perspective on the contemporary relevance of Jewish jurisprudence. It relates some of the dilemmas that confronted the architects of the legal institutions of the State of Israel—among them, Warhaftig—with respect to the question of whether, and how, these institutions should reflect the position of Jewish law on fundamental jurisprudential issues.

One such institution is the rabbinical court; the role of contemporary rabbinical courts in the ongoing formulation of Jewish law is reflected in the chapter by Drori, discussed below. The history of the responsa literature, briefly addressed by Warhaftig in explaining the decision to publish the rulings of the rabbinical courts, is explored in greater detail by Lifshitz, who analyzes that genre in chapter 3.

Another of the legal institutions, the merits of which were much debated, is the doctrine of binding precedent, the main focus of Warhaftig's article. Warhaftig argues that not only is binding precedent not recognized by Jewish law, it is fundamentally at odds with the notion of decision-making underlying Jewish law. While Jewish legal thinking places great weight on the decision-making process, examination reveals that although the importance of judges' being informed by the relevant prior considerations and policies is emphasized, their deliberations and conduct in deciding particular cases are not restricted. They are to exercise judgment independently, basing their decisions on the facts of the cases before them and on the law as they understand it. It is thus the absence of the doctrine of binding precedent, rather than, as might be expected, its endorsement, which attests to the respect accorded the decision-making process in Jewish law. Further, it would appear that to a great extent the perennial vitality of that process can be attributed to the rejection of binding precedent.

Indeed, the very concept of binding precedent is problematic. We are all familiar with the notion that previous cases may be distinguished in a variety of ways. Yet, if a precedent can be ignored by a judge who chooses to distinguish it, in what sense is it binding? Recent studies have suggested that the importance of the doctrine of binding precedent lies in cases where the judge chooses to follow a prior decision, because the court, by citing the doctrine, is able to present its decision as necessitated by pre-existing legal decisions, thereby minimizing criticism and resistance from the losing party. On this understanding, then, the doctrine of binding precedent is construed as a fiction intended to bolster the court's authority.

That Jewish law does not subscribe to the doctrine of binding precedent is particularly significant from this perspective. Apparently, the fact that judges attain their position by dint of their personal moral stature and status as guardians of a valued tradition, and function not only as adjudicators but also as esteemed teachers and communal leaders, obviates the need for jurists to resort to the rhetorical device of binding precedent to advance the authority of the legal system.

The second chapter, by Ben-Menahem, analyzes the status of the judge vis-à-vis the parties and society at large, and the basis of judicial authority. After examining the thesis that the judge's role is modeled upon that of the agent, he argues that this analogy is problematic and should be understood more as a metaphor than as entailing actual legal consequences. In talmudic legal thought, the use of familiar legal institutions to explain more complex relations is not uncommon, despite the fact that such recursive explanation itself raises questions. The phenomenon is best understood in terms of the Talmud's inclination to avoid grounding legal institutions in legislative stipulation (*ex lege*) and, where possible, to prefer instead their derivation from more fundamental legal concepts.

The status of a halakhic decision must be explored not only in terms of its future precedential status, but also with respect to the degree to which it is binding on the parties to the dispute that prompted it. The third chapter, by Lifshitz, considers the distinction between a binding legal decision and a legal opinion issued solely as guidance to the parties involved. This distinction is particularly relevant in analyzing the status of the responsum as a vehicle for legal decision-making. Lifshitz argues that the status accorded the respondents' rulings varied directly with the respondents' authority relative to that of local judges, an interesting example of the interplay between power and process.

Ta-Shma, in the fourth chapter, provides a different perspective on the question of precedent and judicial autonomy, by tracing the historical development and shifts in meaning of the principle that "the law follows the later authority." This principle establishes a central tenet of decision-making in Jewish law.

In the second part of the volume, the first concept examined is that of legal logic, which Ettinger discusses in chapter 5. Ordinarily, logic (*svara*) is conceived of either as a source of law or as a means of interpreting given sources. Ettinger's article, however, explores another use of the concept, prominent in the works of Maimonides, namely, logic as a means of choosing between various sources to decide the law.

Another general jurisprudential concept frequently encountered in Jewish law is that of permissible evasion of the law, examined by Shilo in chapter 6. The pervasiveness of this phenomenon invites speculation as to factors that may have prevented the achievement of the same ends by legislative means. It is possible that the method of permissible evasion of the law was preferred due to ethical considerations. Evasion condoned by the Sages to mitigate the law's harshness in effect generates two levels of conduct: conduct prescribed by legal norms, and conduct sanctioned by permissible evasion of these norms. This bifurcation is not unknown in Jewish law, which establishes an elaborate system of moral aspirations alongside its normative injunctions in order to provide individuals with an opportunity to express their autonomy as moral agents and thereby enhance their moral development. (We have examined this issue in detail in Ben-Menahem and Hecht, *Selected Topics in Jewish Law*, vol. 3, *Law and Equity* (1993).)

Permissible evasion of the law, by providing individuals with an opportunity to forgo evading the law and instead choose to follow the law proper, may constitute yet another avenue for the exercise of such moral autonomy. This idea is nicely captured in the following talmudic passage: "Come and see that the later generations are not like the earlier generations. The earlier generations brought their produce home through the front door in order to be obliged to give tithes. The later generations bring their produce home through roof-tops and courtyards and backyards in order to be exempt from tithes" (bBerakhot 35b).

In the final chapter, Drori explores the subject of inadvertent adultery in Jewish law, touching upon another characteristic feature of Jewish law—recognition of the legal defense of ignorance of the law. The talmudic principle of forewarning stipulates that no offender can be punished until the court is satisfied that the violation was willful. This principle applies to all transgressions of criminal and ritual law. Generally, other legal and social consequences of a violation do not apply if punishment is not meted out due to the offender's ignorance of the law. However, Drori demonstrates that in the case of adultery committed in ignorance of the law, these other consequences do indeed apply. He attempts to explain this unusual treatment of adultery in Jewish law by distinguishing two aspects of criminal offense in Jewish law: the divine aspect, and the societal. While, in the former sphere, ignorance of the law constitutes an acceptable excuse, with regard to acts that have societal repercussions, inadvertent violation due to ignorance is considered tantamount to deliberate transgression, and treated accordingly. That is, ignorance of the law is indeed a legitimate excuse in Jewish law, but only in the realm of the divine. In matters human, in Jewish criminal law, as in Western law, "ignorance of the law is no excuse."

We expect readers to benefit greatly from the scholarship and insight of these studies; we also hope that they will enjoy this encounter with the unique texture of original Hebrew scholarship.

Hanina Ben-Menahem
Jerusalem

Neil S. Hecht
Boston

CHAPTER 1

PRECEDENT IN JEWISH LAW

Zerah Warhaftig

1. Binding and Guiding Precedent

Court rulings, particularly those handed down by supreme courts, can provide invaluable information about a legal system and its implementation in practice. They are a source of instruction, guidance, and perhaps even direction for courts and judges alike. However, the weight to be accorded judicial precedent has been universally problematic since ancient times. What is the status of precedent with regard to similar cases that follow – is it binding or just a guide? Both views can be argued for convincingly.

On the one hand, endowing a legal ruling, particularly that of a supreme court, with binding power, contributes to the unity of law. It ensures a degree of protection against a situation where each court decides idiosyncratically, yielding an unruly miscellany of opinions and principles. The Talmud condemns such a state of affairs: "Do not form separate factions or rival groups."[1] Similarly, endowing precedent with binding force contributes to legal stability, the preservation of legal tradition, and certainty in the interpretation of the law. The parties to a case can predict its outcome with a fair degree of confidence, and individuals can conduct their affairs so as to avoid litigation.

On the other hand, mandating that precedent be followed has many disadvantages. It prolongs or perpetuates judicial error, which tends to entrench itself with each subsequent ruling; it greatly reduces judicial flexibility and devalues the judge's personal outlook on the law and how it should be understood and interpreted against a background of changing circumstances. It detracts from judicial independence of thought and analysis and makes the judge a slave of established patterns: an automaton mechanically deducing foregone conclusions. Binding precedent makes legislators of judges and often causes judges in lower courts to decide against their better judgment and moral intuition. They are obliged to decide the case in accordance with the binding precedent even if it conflicts with their own understanding and interpretation of the law. The judge is thus in effect answerable to two legislators, one of which, the supreme court, in its capacity as supreme legislator, has the final word.

2. The Role of Precedent in Roman Law and Common Law

Precedent was unknown as a source of binding law in Roman law. Judges (*iudices*) were ordinary citizens, generally without any legal training,[2] and their rulings (*iudicata*) were not considered official sources of law.[3] Although Cicero refers to the opinions of legal experts and works on legal matters as binding sources of Roman law, this is not as momentous as it sounds – rulings were binding only for the cases in which they were handed down.[4] Court decisions were not published.[5]

---

1 bJebamot 14a; jPesahim 4:1, 26a, (4:1, 30d), on the basis of exegesis of Deut. 14:1.

2 Y. Pokrovsky, *Toldot Hamishpat Haromai* (Jerusalem: 1928), pp. 114–115; C.K. Allen, *Law in the Making* (Oxford: 1964), pp. 168–169.

3 Justinian, *Institutes*, I. II, 3.

4 Allen, *op. cit.* (note 2 above), pp. 163–164.

5 *Ibid.*, pp. 169–170.

The *Codex Justinianum* (533 CE) explicitly states that courts must not decide on the basis of similar cases, but on the basis of the law,[6] and the *Institutes* cautions judges to decide "on the basis of the law, the constitution and custom."[7]

The jurist Callistratus, quoted in Justinian's *Digest*, reports: "Our emperor Severus ordered that when the law is ambiguous, the prevailing interpretations, as expressed in decisions of recognized authorities on similar matters, acquire the force of law."[8] However, he is the only authority to voice this opinion, and does so with respect to a specific problem, namely, how to interpret an ambiguously-worded law. His solution is to recognize only those interpretations that recur regularly.[9]

Roman law recognizes the importance of learned legal opinion – *responsa prudentium*, and under the rules of Augustus and Hadrian, certain well-known jurists were awarded the *ius respondendi*, or authority to prepare written opinions on cases being tried. These documents, signed and stamped with their authors' seals, were known as *responsa signata*, and the opinions they expressed were binding for the particular cases at hand. The importance of these responsa lies in their influence on legal thought. Some of the official jurists published compilations of their decisions, known as *opiniones* or *sententiae*, which greatly enriched the legal literature.[10]

Most modern legal systems, particularly those based on Roman law, have not adopted the principle of binding precedent to any extent.[11] However, binding precedent does characterize Anglo-American law, derived, as it is,

---

6 *Non exemplis sed legibus iudicandum est. Codex* VII, 45.13.

7 *In primis illud observare debet iudex, ne aliter iudicet, quam legibus, aut constitutionibus aut moribus proditum est* (IV. XVII, 1).

8 *Nam imperator noster Severus rescripsit in ambiguitatibus quae ex legibus proficiscuntur consuetudinem, aut rerum perpetuo similitèr iudicatarum auctoritatem, vim legis optinere debere.* D. 1. 3. 38.

9 Allen, *op. cit.*, pp. 173–174; G. Tedeschi, "Al Hilkhat Hatakdim Hamehayev," in *Mehkarim Bemishpat Artzeinu* (Jerusalem: 1959), p. 100.

10 W.W. Buckland, *A Textbook of Roman Law* (Cambridge: 1932), pp. 21–25; Pokrovsky, *op. cit.* (note 2 above), pp. 145–146; Tedeschi, *op. cit.*, pp. 94–95.

11 H.C. Guterridge, *Comparative Law* (Cambridge: 1946), pp. 113–114; J. Salmond, *Jurisprudence* (London: 1957), pp. 141ff.; Allen, *op. cit.*, pp. 178–186; Tedeschi notes that in three countries, Hungary, Portugal and Argentina, the doctrine of binding precedent is accepted to some degree (*op. cit.*, p. 99). Cf. Civil Appeal 815/75 Leibensohn-Stein et al. v. Authority for the Law for those Disabled through Nazi Persecution, P.D. 32(3), p. 275: "In Germany even a ruling of a high court does not create a binding precedent for that court or for lower courts."

from common law, which was generated from rulings inductively on a case by case basis.[12] In case law of this kind, facts create the law – *ex facto oritur ius*. English law is the model of *stare decisis*,[13] law distinguished by consistent and stable rulings.

The English system spread to all the countries where common law is practiced, and today the principal feature differentiating English and American law from other modern legal systems is the decisive influence of precedent in shaping the law.[14] Cardozo noted that "*stare decisis* is at least the everyday working rule of our law."[15] There has been much discussion recently in the United States about weakening the principle of binding precedent, but it still has a strong hold on the American judicial system.[16]

3. Precedent in Israeli Law

During the British Mandate period, the doctrine of binding precedent was adopted by the courts in Mandatory Palestine as a matter of course, both with regard to precedents set by local courts, and those set by British courts. The principle of binding precedent became part of the legal system by force of article 46 of the Palestine Orders in Council 1922–1947.

Though not without misgivings, this state of affairs continued after the establishment of the State of Israel, until a bench of five judges handed down a supreme court ruling formally adopting the principle of binding precedent in its entirety.[17] This ruling was later interpreted as follows: "It is clear and beyond doubt that the court unconditionally upheld English law in the matter of the binding power of precedent."[18] A lone dissenting opinion, opposed to the absolute nature of the ruling, was rejected.[19]

---

12 Allen, *op. cit.*, pp. 161–162.
13 J. Salmond, *op. cit.*, (note 11 above), esp. pp. 174ff.
14 Edwin W. Patterson, *Jurisprudence* (New York: 1953), p. 300.
15 Benjamin N. Cardozo, *The Nature of the Judicial Process* (New Haven: 1921), p. 20.
16 Edward D. Re, *Stare Decisis*, Federal Judicial Center, (Washington D.C.: 1975).
17 High Court of Justice 287/52; 34/52; 324/52, Ram et al. v. Minister of Finance, P.D. 8, p. 494.
18 Opinion of Justice Vitkon in High Court of Justice 176/54, Nahum Yehoshua v. Board of Appeal for the Law of the Disabled and Others, P.D. 9, p. 624.
19 See High Court of Justice 176/54 cited above; cf. Tedeschi, *op. cit.* (note 9 above), pp. 110, 114, 128.

Afterwards, however, it was decided that the question of the status of the precedent principle should be dealt with legislatively. Clause 41 of the proposed Bill of Law Courts – 1955, declared: "All courts are bound by the rulings of higher courts."[20] Explanatory notes to the clause added: "This law does not settle in principle the issue of binding precedent, i.e., the question of the extent to which a court – and particularly the Supreme Court – is bound by its own prior decisions. This question will remain legislatively unresolved, and its solution is to be sought in the realm of the judiciary; nevertheless, the rule that a lower court is bound by the decision of a higher court has gained such a degree of acceptance, and is so established today, that it is possible, and desirable, to promulgate it as law."[21]

Debate on the proposed law began in the second Knesset,[22] but it did not receive a second and third reading. The Bill of Law Courts was put forward again at the third Knesset,[23] with some modifications.[24]

In the ensuing debate, in November 1955, I opposed the clause, on the grounds that the notion of binding precedent was an English concept that we had adopted despite the fact that it was incompatible with the principles underlying traditional Jewish legal thought. I explained that to my mind binding precedent was an undesirable means of turning judges into lawmakers. Further, I argued that it might induce judges to circumvent existing law. I also expressed concern about binding precedent in view of the fact that the courts' purview at times extends to adjudication of cases involving questions of halakha. Erroneous rulings in such cases would be tantamount to erroneous halakhic rulings, which, in turn, would become binding precedent for other courts. I suggested that it would be better to view precedent as guiding, rather than determining, future rulings.[25]

---

20 *Proposed Law* 226, 1956, p. 70.

21 *Ibid.*, p. 70.

22 *Divrei Haknesset Hashnia*, vol. 17, pp. 709ff, 718ff, 736ff.

23 *Divrei Haknesset Hashlishit*, vol. 19, p. 316. Although the proposal was tabled on October 12, 1950, a different date is given for the event on p. 289 (November 14, 1955).

24 Some of the modifications were noted by Minister of Justice P. Rosen, *ibid.*, p. 329.

25 *Ibid.*, p. 322; cf. remarks by Knesset Member Zalman Ben-Yaakov in *Divrei Haknesset Hashnia*, vol. 17, p. 722.

The Knesset legal committee suggested that some qualifications regarding the power of precedent were indeed called for and recommended that clause 33 of the proposed law read as follows:

a. Courts will be guided by rulings of higher courts.
b. Rulings of the Supreme Court are binding on all courts other than the Supreme Court.
c. Rulings of the Supreme Court handed down by a bench of five judges are also binding on the Supreme Court.

I expressed my reservations about the proposal on the Knesset floor, and suggested it be qualified in two ways. First, I argued that clause 33 should be omitted, for a number of reasons. I reiterated my objections to the doctrine of binding precedent: it might perpetuate erroneous rulings; mislead judges into false analogies; reduce the flexibility of legal decisions in the context of changes in commerce and business arrangements; and discourage independent legal reasoning on the part of judges. Without denying the advantages of the precedent system, I noted that the whole matter of the role of precedent should be left to the discretion of the courts rather than defined by legislation. Legislating the status of precedent amounts to endorsing one ruling for all similar cases, effectively making a legislator of the judge who handed it down. This, in turn, would undermine the division of authority between the judicial and legislative branches and could ultimately damage public confidence in adjudication. I went on as follows:

> Whatever our position on the question of binding precedent, one thing is clear: determination of the status of precedent is a concern of the courts. Just as one should not confuse the written word with the spoken word, so too there should be no confusion between law and legal precedent. About a year ago, the Supreme Court ruled that its decisions are binding, not only on lower courts, but also on the Supreme Court itself. This ruling does not worry me, because it is conceivable that though the Supreme Court indeed ruled in favour of binding precedent, it may at some point in the future be persuaded of its error in so doing. In any event, this is possible only as long as it falls under the courts' purview. But if we pass a law that rulings of the Supreme Court are binding both on lower courts and on the Supreme Court itself, we are in effect transforming the rulings of the Supreme Court into law. If this bill becomes law, should the Supreme Court decide an issue erroneously, its ruling will nevertheless be law, and there will be no remedy but to turn to the legislature and go through the process of changing it.
>
> This would turn the Supreme Court into a second legislative assembly. I do not think it advisable for the Knesset to relinquish its legislative prerogative. The clause, as it stands, delegates a unique legislative power to the Supreme Court,

giving it the power to change any law enacted by the Knesset simply by interpreting it incorrectly, even if the error is obvious. I believe that this clause, if passed, will petrify precedent and render a disservice to our legislation.[26]

Together with Y. Harari, I also proposed deletion of the third sub-clause, (c), on the grounds that the Supreme Court's authority to render an independent decision should not be limited in any way.[27]

My first qualification – omission of section 33 – was rejected.[28] On the other hand, the Warhaftig-Harari amendment on sub-section (c), was passed.[29] Clause 33 of the Law of Courts 5717–1957, became law. It reads:

a. Courts will be guided by rulings of higher courts.
b. Rulings of the Supreme Court are binding on all courts other than the Supreme Court.[30]

The scope of binding precedent as understood and applied by the courts up to that point – as is clearly reflected in the decision of the Supreme Court in the *Ram* case[31] – was thus significantly reduced by the final version of clause 33, which freed the Supreme Court from the obligation to rule in accord with its previous decisions.

The main contribution and innovation[32] of clause 33 lies in the fact that the binding status of precedent, and the very significant limitation on it, is articulated legislatively. The legislature limited its own purview by extending that of the Supreme Court, which was granted tremendous legislative authority. Clause 33 also establishes two classes of judges: the few who have the authority to interpret laws and render judgment independently, and all the

26 *Divrei Haknesset Hashlishit*, vol. 22, p. 2492.
27 *Ibid.*, pp. 2495–2497.
28 *Ibid.*, p. 2498. The first vote was a 16 : 16 tie, and a majority of 18 : 16 against the amendment was reached only on the second vote.
29 *Ibid.*, p. 2497.
30 *Sefer Hahukim* 233, 1957, p. 152.
31 See note 17 above.
32 In England and the United States *stare decisis* is based on court rulings, and has no basis in the constitution or in law. In Portugal and a few Latin American countries, such as Argentina, legislation or the constitution is used to establish that courts are bound by supreme court decisions. However, this dependence usually obtains only in cases where the ruling recurs a number of times. In Mexico, for example, it comes into force only if a ruling has been handed down five times. See Tedeschi, *op. cit.* (note 9 above), p. 99.

rest, whose authority to engage in independent legal analysis is restricted by clear subordination to the rulings of the Supreme Court.[33] The sole authority they retain is the authority to compare and contrast different cases – to decide whether or not the matter before them is comparable to a situation covered by an existing decision of the Supreme Court.

The law was received by the judiciary with doubt and apprehension, particularly on the part of judges presiding over district courts.[34] The justices of the Supreme Court, on the other hand, generally greeted it with understanding and approval. This attitude is reflected in *Lapidot Israel Petrol Co. Ltd. and Others* v. *Shlisser*.[35] The bench of five justices declared that the Supreme Court, though not legally bound by precedent, was not in the habit of deviating from earlier decisions; however, with regard to any ruling issued routinely, without deliberation, it was the Supreme Court's responsibility to evaluate whether it was proper to continue in this direction in the future.

The Supreme Court chose moderation with regard to reversing rulings it had handed down previously, arguing that judicial tradition and legal stability ought to be maintained. Justice M. Silberg declared candidly, "I do not disregard the decisions of earlier authorities."[36] At the same time, the Supreme Court does not hesitate to deviate from an earlier ruling it regards as erroneous.[37]

Justice M. Landau found it objectionable that the status of precedent in the legal system was being defined legislatively – "Personally, I find Clause 33 of the law courts bill altogether unnecessary."[38]

---

33 See A. Hecht, "Hazika Letakdim", in *Mehkarei Mishpat Lezekher Avraham Rosenthal* (Jerusalem: 1974), pp. 162, 168, 175.

34 See G. Tedeschi, "Olelot Bedivrei Hatakdimim," *Hapraklit* 17 (1961), p. 244; Hecht, *op. cit.*, p. 172.

35 Additional Hearing 14.68, P.D. 23(1), p. 771.

36 Civil Appeal 476/65, Rinat v. Rinat, P.D. 20(2), p. 25, under "Alef."

37 Civil Appeals 606/73, 126/74, Yardenia Ltd. v. Ofer Brothers Ltd.; Port Authority v. Ararat Ltd., P.D. 30(8). The comment of Justice Etzioni is noteworthy: "I have the honor of being the first to deviate from English law, according to which, in matters of damages, what counts is the date damages were inflicted rather than the date of sentencing" (p. 42, under "Alef"). He then quotes from a ruling he himself handed down as a district court judge.

38 M. Landau, "Halakha Veshikul Daat Beasiyat Mishpat," *Mishpatim* 1 (1969), p. 295. Landau's view is as follows: "If that which may be termed 'judicial legislation' indeed exists, then in functioning as secondary legislatures, the courts should enjoy autonomy and the freedom to define for themselves the manner in which they will legislate, for they are responsible for the adequacy of the legislation thus generated" (*ibid.*).

Justice H. Cohn mentioned a number of principles that inform Supreme Court policy on earlier rulings. The accepted practice is not to utilize the license to deviate from precedent, even if the court disapproves of the ruling as it stands. In Cohn's opinion, however, this policy applies chiefly to precedents pertaining to the determination of judicial jurisdiction or procedure, areas in which stability and certainty are of utmost importance. "If deviating from precedent is likely to obstruct justice, and any benefit accruing from it will be purely academic, it is preferable to stay on the beaten track of precedent." However, he goes on to say that it is quite another matter if the court is of the opinion that a past decision could bring about injustice, either because circumstances have changed, or because it was mistaken from the start.[39] To support his position, Cohn cites several sources from Jewish law, including *Piskei Harosh*.[40] These remarks echo views expressed in a lecture entitled "The Problem of Precedent in Jewish Law," which Cohn delivered at the College of Law and Economics in January 1957.[41]

In support of my reservations about clause 33 during the Knesset debate, I too appealed to Jewish law. The remainder of this article will consider the position of Jewish law on the question of the role of precedent.

4. Previous Decisions and Precedent in Jewish Law

i. The Significance of Actual Rulings

Jewish law is basically statutory law. It is regarded as divine in origin – "For the judgment is God's";[42] "For all His ways are justice."[43] Man does not create the law; he reveals it. The task of earthly judges, in the words of Isaiah, is to "learn to do well; seek justice,"[44] i.e., to study the sources, discover the fundamental principles underlying them, and decide in accordance with them, in order to administer "justice and righteousness."[45] The purpose

---

39 Additional Hearings 39/75, 41/75, Port Authority v. Ararat Ltd. et al.; Ofer Brothers v. Port Authority et al., P.D. 31(1), p. 535. See H. Avnor, "Sidrei Hadin: Hatzedek Vehamediniyut," *Hapraklit* 32 (1980), p. 478.

40 Cohn, *op. cit.*, p. 537, citing *Piskei Harosh*, Sanhedrin 4:6.

41 Published in *Mishpat Vekalkala* 3, 3 (March 1957), pp. 129–141. The Rosh is cited on p. 136.

42 Deut. 1:17.

43 Deut. 32:4.

44 Isaiah 1:17.

45 I Kings 10:9.

of the law is to administer justice and equity on earth,[46] "for right shall return unto justice."[47]

Divine law was handed down as the written law, the Torah, but "all the laws given to Moses on Sinai were given with their explanations ... and Moses did not commit to writing the commandments which are an explanation of the Torah, but charged [them] to the elders and to Joshua and to the rest of the nation orally ... and for this reason they are called the oral law ... and it was repeated to the Sages publicly and revealed to all Israel and they committed it to writing."[48]

The solution to every problem should be sought, and can indeed be discovered, in the written and oral law, the Mishnah and the Tosefta, the Midreshei Halakha, the Babylonian Talmud and Jerusalem Talmud, and the halakhic literature.

Unlike Roman law, Jewish law attributes great significance to judicial facts: previous decisions handed down on a given issue. The actual implementation of the law leads to its theoretical clarification. Indeed, the talmudic literature takes note of the critical importance of actual rulings and their implementation: an actual ruling is authoritative;[49] an actual ruling is more weighty;[50] application of the law is greater than studying the law;[51] what has been done is no longer open to discussion;[52] the Elder [R. Jose] has already ruled;[53] had the case come before him he would have been more punctilious;[54] he saw the act and recalled the law;[55] did you ever rule on such a case?;[56] behold day-to-day practice;[57] we learn from everyday practice.[58, 59]

46 Jeremiah 9:23.

47 Psalms 94:16.

48 From Maimonides' introduction to the *Code*.

49 bShabbat 22a, Rashi *ad loc.* s.v. *maase rav*; bBaba Batra 130b; bNida 65b; cf. Rabbenu Hananel on Shabbat 49b, and Tosafot *ad loc.* s.v. *aba salha.*

50 bBaba Kama 83a.

51 bBerakhot 7b, bTemura 16a.

52 bRosh Hashana 29b.

53 bShabbat 51a, bJebamot 105b, bSanhedrin 24a, 29b.

54 Rashbam, bBaba Batra 83a s.v. *umar savar maase rav.*

55 bPesahim 66a.

56 bBeitza 9b.

57 bSuka 44a; bKetubot 68b, 95b, Rashi *ad loc.* s.v. *veha maasim bekhol yom*; bBaba Kama 95b; bBaba Batra 173b.

58 *Tosafot*, Shabbat 48a s.v. *mai shna mimeiham.*

59 For more on actual rulings in Jewish law, see M. Elon, *Jewish Law*, (Philadelphia and Jerusalem: 1994), vol. 2, pp. 945ff.

An important feature of the practice of law is judicial expertise. Unlike the Roman system, in which laymen acted as judges, Jewish law mandated that judges be trained and officially authorized legal experts,[60] for, as administrators of divine law, they had to be masters of their craft.[61]

ii. Publication of Responsa

There are no official compilations of court rulings or proceedings in Jewish law, but case histories and decisions are found informally in both the Jerusalem Talmud and the Babylonian, the Midreshei Halakha, Tosefta, and other halakhic works.[62] The thousands of volumes of published responsa are a veritable treasure trove of halakhic rulings.[63] These are sometimes indexed, and recently several scientific indexes to the responsa literature have been compiled.[64] One hundred and fifty volumes of responsa, containing thirty-five thousand individual rulings, have been computerized, facilitating speedy access to the judicial decisions therein.[65]

These works contain responses on a wide variety of subjects, and their nature and style vary with the period during which they were written. Collections of responsa from the Geonic period generally contain legal rulings with only the briefest of explanations, and sometimes with no explanation at all. The responsa of the Early Authorities are more detailed, while many responsa of the Later Authorities include extensive explication and discussion of the laws in question. Often, responsa are written by rabbis and community leaders regarding cases whose litigants did not appear before them; such responsa tend to be expository in nature, focusing on clarification

60 Deut. 16:18, Exod. 22:7–8, bSanhedrin 56b.

61 Exod. 18:21; bSanhedrin 2a, 5a; jSanhedrin 1:1; Maimonides, *Code*, Laws Concerning the Sanhedrin, 1,2; Tur and *Shulhan Arukh*, Hoshen Mishpat, secs. 1,2.

62 Elon, *loc. cit.* (see note 59); E.Z. Melamed, "Hamaase Bemishna Kemakor Lehalakha," *Sinai* 46 (1960), pp. 152–166.

63 Cohen, B. *Kuntras Hateshuvot* (Budapest: 1930; reprinted Jerusalem: 1970); Elon, *op. cit.*, vol. 3, pp. 1453–1528.

64 The Hebrew University's Institute for Research in Jewish law published a detailed index to the responsa of R. Asher b. Jehiel in 1970, and subsequently published legal indexes and source indexes to the responsa of Spain and north Africa. A number of indexes have also been published by Bar Ilan University's Institute for Research on Oriental Jewry.

65 The responsa project was carried out by the Institute for Data Processing and Cognitive Linguistics at Bar Ilan University.

of the relevant laws, rather than rulings proper. Sometimes the writer emphasizes that his work is only intended as a theoretical elucidation of the matter and should not be construed as a decision intended for implementation.[66, 67]

Many responsa deal with actual cases which were adjudicated by the respondent, but written up later; they do not record the decisions as handed down at the time of the adjudication, but instead analyze the cases retrospectively for future reference. However, there are also collections of responsa containing rulings signed by courts of three judges.[68] Other rulings handed down by such courts are preserved in community records.[69] Finally, expository works on halakhic matters sometimes formulate hypothetical questions on thorny points, then proceed to answer them.[70]

iii. Publication of Rabbinical Court Decisions in Israel

An important innovation in the history of the responsa literature was inaugurated in Israel with the decision to publish the rulings of the Rabbinical High Court of Appeal and those of the district rabbinical courts. The rulings are published together with the arguments on which they are based, as presented in court. Indeed, I myself proposed the publication project, and was charged with its implementation, a responsibility I viewed as a great privilege and sacred trust.

Previously, the Rabbinical High Court of Appeal followed the traditional system of issuing brief rulings while at the same time compiling a full account of the halakhic deliberation on the case in pamphlet form for circulation among judges. Deliberation and discussion are an essential part of the legal process, allowing the individual judges an opportunity to convince their colleagues of the validity of their arguments, so that a decision can be

---

66 Rashba, *Responsa Attributed to the Ramban* #112, #222; Tashbetz, part 1, #85, part 2, #5; and many like these.

67 See E. Shochetman, "Hovat Hahanmaka Bemishpat Haivri," *Shenaton Hamishpat Haivri* 6–7 (1979–80).

68 E.g., R. Samuel Di Modena (Maharashdam), *Responsa*, Even Haezer #27, #43, #60; R. Isaiah Basan, *Lahmei Toda* #31.

69 Israel Heilpren, *Pinkas Vaad Arba Aratzot* (Jerusalem: 1945), see index listing rulings, pp. 567–568; Moetzet Harabanim Bemorocco, *Haaseifa Hashnatit Hashlishit* (Proceedings of the Rabbinical Council of Morocco), (1950), pp. 30–31, and many others.

70 R. Israel Isserlein, *Terumat Hadeshen*, part 2, Pesakim Uktavim, quoted in R. Shabbetai Cohen (Shakh) on the *Shulhan Arukh*, Yore Deia, sec. 196, par. 20.

reached. The pamphlets were intended to facilitate this process, rather than to explain the rulings to the litigants involved, so that they could understand why they had won, or lost, their cases.[71] There was no appeal against a ruling of the Rabbinical High Court, nor were there established procedures for appealing the rulings of district rabbinical courts. (Interestingly, these pamphlets often served as the basis for volumes of responsa published by their authors years later.[72])

The idea of publishing, in an organized fashion, both the courts' rulings and their grounds, and that of appending abstracts of the laws cited in the rulings, as is customary in law reports, to allow for ease of reference and study, was thus entirely new. Accordingly, the Chief Rabbinate, which had to approve the proposal, had to be convinced of its merits. This entailed some negotiation, in which, as head of the Ministry of Justice's Research Institute for Jewish Law, I was much involved.

In due course an agreement in principle was reached between myself and the Chief Rabbinate.[73] After some administrative changes were carried

---

71 Cf. bSanhedrin 33b. Also see Shochetman, *loc. cit.*

72 E.g., the responsa of R.A.I. Kook, *Ezrat Kohen* – Teshuvot Behilkhot Even Haezer (Jerusalem: 1969) and *Orah Mishpat* (Jerusalem: 1979); and the following responsa: R. Yosef Zvi Halevi, *Amira Neima*; R. Benzion Hai Uziel, *Mishpetei Uziel*; R.I. Herzog, *Heikhal Yitzhak*; R.A. Hadaya, *Yaskil Avdi*; R.M. Roth, *Kol Mevaser*; R. Ben-Menahem, *Divrei Menahem*; R.E. Waldenberg, *Tzitz Eliezer*, R. Ovadia Yosef, *Yabia Omer*; R. Werner, *Mishpetei Shmuel*; R.Y. Aaronow, *Divrei Yehoshua*; and others.

73 Due to its historical importance, it is appropriate to quote the full text of the agreement between the Chief Rabbinate and myself:

> A special session at the home of R.I. Herzog, under the chairmanship of Chief Rabbis I. Herzog and Benzion Uziel, held on Sunday, the third day of Adar 1948. Those present: Dr. Zerah Warhaftig, R. Dr. S.Z. Cahana, Dr. S.B. Feldman, R. Weber, first secretary, and M. Lebanon, legal advisor.
>
> Dr. Warhaftig has proposed publishing the rulings of the rabbinical courts under the editorship of R.S.Z. Cahana and Dr. Feldman with the assistance of the first secretary, R. Weber, and the legal advisor, Mr. Lebanon.
>
> R. Weber pointed out the difficulties in the publication of rulings the rationales for which have thus far only been articulated in lengthy pamphlets. Some rabbinical judges involved object to publication because their rulings are not immediately comprehensible; furthermore, publication of the rulings would require a standardized format, something outside the province of the editors. Mr. Lebanon suggested that before discussing

out,[74] the first collection of rulings of the Chief Rabbinate's Rabbinical High Court of Appeal was finally published in 1950. Assisted by S.B. Feldman, S.Z. Cahana, and P. Galevsky, I served as editor. In the foreword to the volume, I wrote:

> The selection of the rulings herein published was guided by the desire to accurately portray the workings of the Court. Most of the rulings relate to family law and public endowments; the others are devoted to monetary matters. The opinions of the judges, with a few exceptions, are not published as written, but have been abstracted by the editors from the contents of the pamphlets appended to the case files. This volume thus does not constitute a formal record and the editors assume full responsibility for the adaptation and wording of the judicial opinions.

The collection contains fifty rulings on matters of consequence, set out and worded as clearly as possible. The principal facts are presented succinctly, the conclusions unambiguously. It thus provides jurists and scholars with access to the world of the rabbinical courts and the methods by which they reach their decisions.

It was found that publication encouraged rabbinical court judges to communicate their opinions in a clear and orderly manner comprehensible to those unschooled in Jewish law, whether jurists or members of the public. Over time, rulings of the Rabbinical High Court of Appeal and the district rabbinical courts began to be handed down in a form that allowed them to be

---

adaptation of the pamphlets, the proposed editors familiarize themselves with the material by reading a number of pamphlets.

Dr. S. B. Feldman stated that it appeared he would have to edit the factual material and R. Cahana, the halakhic material.

Dr. Warhaftig pointed out that this would not be an official report, hence there was no need to be too pedantic, and that only a selection of cases would be published. Rulings that could be omitted would not appear. He suggested that a number of files be provided to R. Cahana and Dr. Feldman; and that once they had considered them, the five of them could reconvene and proceed from there.

The proposal was accepted and it was decided to hold another meeting the following Tuesday at the office of first secretary R. Weber.

Signed, I. Halevi Herzog, Benzion Meir Hai Uziel.

74 The first secretary of the Chief Rabbinate, R. Weber (Shazori) of blessed memory, and the legal advisor, M. Lebanon, withdrew from the project, apparently believing it would not succeed.

published as written, with no editing. Accordingly, it was decided to publish the rulings of the district rabbinical courts, and later, those of the Rabbinical High Court of Appeal, on a monthly basis.

In my foreword to the first volume of rulings of the district rabbinical courts, I remarked on its significance:

> This first volume of rulings of the district rabbinical courts marks the inauguration of a series to be published regularly. From the perspective of halakhic as well as legal literature, it constitutes an important innovation. Hitherto, the custom of rabbinical courts was to produce the reasons for their rulings only if so requested by a litigant who wished to know the basis for the ruling. However, in the Diaspora, when rabbinical courts had no means of enforcing their decisions, and their authority rested solely on the voluntary consent of the litigants, such explanations were usually unnecessary. In most countries there was no hierarchy of rabbinical courts – all rabbinical courts were at the same level – so that there was no possibility of appealing a decision by petitioning a higher court. The need to be apprised of the grounds for the rulings thus arose solely from scholarly considerations. The rabbis and judges, of course, deliberated before reaching a decision, and often composed written reports of their opinions, but these were only rarely published. Even when they were published, in the form of volumes of responsa, this was often many years after the rulings had been handed down. Overall, however, only a small fraction of rabbinical court rulings were indeed published. These rulings, therefore, while of value for scholars, were of no benefit to the parties involved in the cases.
>
> With the rebirth of rabbinical courts in Israel ... a major change has taken place. Rabbinical courts now have the authority to enforce their rulings. An appellate court has been established, making it possible for rulings to be referred to a higher level. Familiarity with the grounds for a given ruling thus has practical value: it enables litigants to decide whether, and on what grounds, to appeal a lower court ruling. The Rabbinical High Court of Appeal, too, must be conversant with the reasons for a ruling under review if it is to clarify the laws in question and decide correctly. Furthermore, an understanding of the legal reasoning involved allows lawyers to better handle similar cases in the future. Publication of the rulings provides them with direct access to the pertinent talmudic sources. According to the law, a rabbinical court with the authority to enforce its decisions must give the reasons for these decisions in writing (*Shulhan Arukh*, Hoshen Mishpat, sec. 14, par. 4, according to R. Joseph Caro and R. Moses Isserles; Caro's *Avkat Rokhel*, 17, 18, 19; *Noda Biyehuda* on Hoshen Mishpat, 2nd ed., sec. 1). On the basis of this principle, the Chief Rabbinate directed all rabbinical courts in Israel to tender their opinions in writing (*Takanot Hadiyun* 1943, sec. 80).
>
> As an affiliate of the Chief Rabbinate, the Rabbinical High Court of Appeal takes this directive very seriously, to the point of sending files back for revision

> if the judicial reasoning is inadequately articulated...(see *Rulings of the Rabbinical High Court of Appeal*, ed. Z. Warhaftig, p. 153[75]).[76]

In addition to the inaugural volume of rulings of the Rabbinical High Court of Appeal, eleven volumes of rulings of Israel's rabbinical courts had been published by 1960. These well-indexed[77] volumes alone contain a wealth of decisions on questions of family and monetary law and on matters of vital public interest.

5. Precedent in Jewish Law – Guiding Precedent

Rulings, both those in the responsa literature and those published on a regular basis, are invaluable tools for the study of Jewish law. Although courts are not bound by earlier rulings, previous decisions shed much light on how its sources are to be approached and interpreted. The role of the judge is to hand down a true and inherently just judgment,[78] to the best of his ability, accepting full responsibility for the decision he hands down. He may – indeed, he is obliged to – consult those whose knowledge is greater than his own, study the literature, and seek precedents, but the decision must be based on his own best judgment.

---

75 For a recent example, see Appeal 137/1977, *Rabbinical Court Judgments* [henceforth, RCJ], vol. 10, p. 327. On p. 331, the Rabbinical High Court declares, with regard to the majority view of the Ashdod rabbinical court cited there (case 355/1977, p. 318):

> Therefore it seems that the truth is with the minority opinion that the appellant does not need a bill of divorce at all, not even *lehumra* [to satisfy the scrupulous]; as opposed to the majority view of the district rabbinical court to the effect that a bill of divorce is necessary *lehumra*, **because it gave no halakhic reason for its ruling, and this is unacceptable** (emphasis in the original).

76 RCJ, vol. 1, ed. D. Katz and Y. Glasner, p. 3; cf. the address by the Chief Rabbi of Israel, President of the Rabbinical High Court, R.I. Halevi Herzog, p. 2.

77 A combined index to the first five volumes of RCJ appeared in 1967, edited by R.D. Katz. Publication of a combined index, including an index of sources, for all rabbinical rulings to date, should be considered.

78 bSanhedrin 7a, 111b; bBaba Batra 8b; *Tosafot ibid.* s.v. *din emet leamito*; bShabbat, 10a; bMegila 15b.

In Jewish law, then, precedent guides the judge, but the judge is not obligated to follow it: it is not binding. The principle is: "all agree that we do not learn the law from a previous ruling";[79] "if, say, one's teacher ruled on a point of law, one should not determine the law accordingly, because one might be wrong about his teacher's reasons for that particular decision, since there are many who err in their studies."[80] Thus the law must not be decided on the basis of a previous decision; previous decisions are to be used only as guidelines.

The weight accorded a given precedent is influenced by the stature and authority of the court which sets it, on the one hand, and by the stature, authority, logic and halakhic insight of the judge who employs it, on the other.

R. Joseph ibn Migash, in the late twelfth century, was asked whether one who "understands neither the essence of a law nor how it is derived from the Talmud" may rule on the strength of Geonic responsa, since "in their responsa, many Geonim ruled on a matter ... but later they changed their minds or were criticized by others." He replied as follows:

> One who does not rely on himself, but follows the rulings of the Geonim, which are decided laws (*halakhot psukot*), given with succinct and clear reasoning, such a person is to be commended more than those who presume to rule on the basis of the Talmud ... because there is no one in our generation whose mastery of Talmud is such as to enable him to arrive at a decision without considering Geonic opinions.[81]

About two hundred years later, R. Jacob Halevi Molln (Maharil) writes to R.A. Katz:

> As to your remark that responsa should not be relied upon, on the contrary, they are rulings intended for implementation and therefore we learn more from them than from the works of the authorities which are not actual rulings ... hence the widely-used expression "regarding the law as put into practice" ... . And as all our masters from the post-talmudic period quote from the responsa of their predecessors, should not we, who are but orphans of orphans, do likewise?[82]

Overdependence on precedent is a sign of weakness: as R. Joseph ibn Migash says, "there is no one in our generation whose mastery of Talmud is such as to

79 jHagiga 1:8, 7b. This dictum was spoken by R. Hanina in the name of Samuel.

80 *Korban Haeida ad loc.* s.v. *sheein lemidin min hamaase.*

81 *Responsa of R. Joseph Migash*, #114. On this responsum cf. Elon, *op. cit.*, vol. 3, pp. 1181–1184; Ta-Shma, "Yetzirato Hasifrutit Shel Rabeinu Yosef Halevi Ibn Migash," *Kiryat Sefer* 40 (1971), pp. 547ff.

82 Maharil, *Responsa*, #72; cf. Elon, *op. cit.*, vol. 2, pp. 976–977.

enable him to arrive at a decision";[83] and as the Maharil says, we are "orphans of orphans."[84] R. Abraham, Maimonides' son, was critical of this point of view:

> In my opinion a judge who decides only on the basis of written and explicit works, is weak and feeble ... this is not right; these written works are indeed cardinal, but the judge ... has to consider them with reference to the case before him, and apply the laws to similar cases, so as to arrive at conclusions regarding the case at hand on the basis of these cardinal principles. ... and the numerous incidents we find in the Talmud which mention specific laws are not related gratuitously, or to enable judges to decide contemporary cases according to what is said there, but rather, to endow the wise, upon hearing these incidents many times, with the good judgment and analytic skill to hand down proper rulings.[85]

R. Judah, the son of R. Asher b. Jehiel, notes that:

> Nowadays, given that we have manuals of the law [he is apparently referring to *Piskei Harosh*] we ought not rely on responsa. And if I find that such works and the responsa contradict each other, I decide according to the authorities who wrote the manuals, because they are later.[86]

To conclude, let me quote from R. Yomtov of Seville, the Ritba,[87] of the thirteenth century, one of the Early Authorities, quoting his teacher R. Aaron Halevi:

> [A previous ruling is binding] only for the given case, but in a different case, even though it is exactly like the first one, a judge is bound only by what he beholds with his own eyes.

6. Arguments Against Implementation of the Doctrine of Binding Precedent in Jewish Law

There are three main halakhic objections to the doctrine of binding precedent.

i. Apprehension that judges will err by comparing unlike cases.
ii. Judges are not bound by earlier rulings, but are independent and may rule differently.

---

83 See text above at note 81.

84 See above.

85 R. Abraham, son of Maimonides, *Responsa*, #97. Cf. Elon, *op. cit.*, vol. 2., p. 968.

86 *Zikhron Yehuda* #15.

87 Ritba on Baba Batra 130b; also cited in *Hidushei Haran* on Baba Batra *ad loc.*, and *Nimukei Yosef* on Alfasi *ad loc.*; cf. Elon, *op. cit.*, vol. 2, p. 983.

iii. Judges are obliged to decide the cases before them and cannot simply reproduce decisions handed down by other judges.

Let us examine each argument and its basis in Jewish law.

i. Apprehension About Error

Analogy is an important tool that tremendously enriches both legal thought in general and Jewish law in particular. "Since the Torah is brief in one place and expansive at another,"[88] it stimulates legal thought and trains the judge to find innovative solutions to the novel problems that are constantly surfacing. "It is well known that countless new cases arise with the passage of time, and they cannot all be committed to writing along with explication of the relevant laws."[89]

In rendering judicial decisions, it is often the case that analogy is used; in civil law, liability can be established on the basis of judicial reasoning.[90] The Talmud notes: "In all legal matters we compare things to each other."[91] This is explained by the *Yad Rama* as follows:

> In all legal matters (except the laws of forbidden food) we compare things, because not all the laws and all the particulars of prohibitions are specified explicitly. If a matter arises which is not written explicitly, we have to compare it with another which is similar from among those which explicitly appear in the Talmud. And we see that the Sages of the Gemara in all their teaching and in ruling on cases, learned one thing from another and compared one thing with another.[92]

R. Solomon b. Simon Duran expressed this idea in verse:

> For the wise rely
> on that elsewhere spoken
> recognized by all,
> and handed down as well,
> to enlighten both simpleton and scholar.[93]

But this penchant for analogy can have a negative impact on the mental acuity required to distinguish between different facts and laws, encouraging error, banality and vapidity. "R. Joshua b. Levi compares one thing with another, but

---

88 jRosh Hashana 3:5.

89 *Zikhron Yehuda* #79.

90 *Tosafot*, Baba Kama 4b s.v. *veeidim zomemin*; *ibid.* 2a s.v. *velo ze vaze.*

91 bBaba Batra 130b.

92 *Yad Rama ad loc.*

93 Rashbash, *Responsa*, #1.

R. Eliezer does not."[94] "Should we act just because we compare the things?"[95] Great care must be taken in comparing one case to another, given the ever-present possibility of mistaken analogy.

The Gemara censures the judge who "has a case brought before him and decides the law by comparing one thing to another, yet he has a teacher and does not consult him."[96] Every case must be judged independently; hence the judge who is inflexible, who is determined to apply a given law and searches for precedents to that end, deserves censure. The Tosefta remarks in the same spirit, "we do not question a ruling on the basis of another ruling issued by one who is fixated on a precedent."[97] Maimonides cites this principle as law:

> Any judge who has a case brought before him and begins by comparing it with cases in which rulings have already been handed down, and, though there is an authority wiser than himself in his country, does not consult him, such a judge is one of the malefactors who administer justice contemptibly.[98]

The *Shulhan Arukh* states:

> One who has a case brought before him and compares it to another case which he ruled on previously, and, though there is in the city one wiser than himself, does not consult him, such a judge is one of the evil who administer justice contemptibly.[99]

Although in the *Shulhan Arukh* the phrase used is 'which he ruled on previously,' that is, it refers to a judge ruling in accordance with a precedent he himself set in the past, Maimonides' version clearly applies to any precedent, whether set by the judge in question or by another court. This is clear from the words of the Gemara quoted above "and decides the law by comparing one thing to another."[100] The judge decides the law on the basis of another case, not necessarily a judgment he handed down himself.

---

94 bBerakhot 19a.

95 bGitin 19a.

96 bJebamot 109b.

97 Tosefta Kilayim 5:4, cf. commentary of Rash *ad loc.*; cf. S. Lieberman, *Tosefta Kifshuta*, Seder Zeràim, part 2, p. 651.

98 Maimonides, *Code*, Laws Concerning the Sanhedrin 20:8.

99 *Shulhan Arukh*, Hoshen Mishpat 10:2; this also appears in the Tur, *ibid.*, sec. 3, with slight changes in the wording.

100 See text above at note 96.

Ketzot Hahoshen[101] tries to limit this rule to cases in which the judge does not consult anyone at all; once he has consulted the greatest authority in his city, however, he need not confer with him again regarding a similar case. *Ketzot Hahoshen* is apparently following the Rashbam: "He should act all his days according to this and he need not ask about each particular case which comes before him."[102, 103] But in his *Bayit Hadash*, R. Joel Sirkis differs: "He must confer with his teacher on every case."[104] The *Prisha* explains this view as follows: "He has to consult his teacher repeatedly, lest he err about the second case being completely analogous to the first."[105, 106] The interpretations of R. Joel Sirkis and the *Prisha* are supported by Rashi: "The judge decides the law: he knows the law, but it does not really cover the case at hand, and he compares one thing with another and does not consult his teacher."[107] Since the principal reason for condemning the judge who "sticks like glue to a rule" is the possibility of his making a false analogy, and each case must be treated as one that is entirely new, if it is incumbent upon him to consult his teacher or a local authority greater than himself, this obligation obtains regarding each and every case he adjudicates.[108] Generally the facts, the contested issues, and the financial implications differ from case to case, though they may be similar. Any one of these differences may influence the outcome of the case, thus any comparison between cases entails intense scrutiny of the details involved so that relevant differences do not escape notice. The danger inherent in facile analogies is clear.

There are many cases in the Gemara in which attempts to use a previous ruling as the basis for deciding an ongoing case fail when, upon examination, the initial judgments are seen to follow from some specific detail which affected the outcome of the case. Let us examine a few such cases.

---

101 *Ketzot Hahoshen* on Hoshen Mishpat, sec. 14, par. 1.

102 Rashbam, Baba Batra 130b s.v. *yelekh veyaase maase*.

103 *Nimukei Yosef ad loc.* concurs.

104 *Bayit Hadash* commentary on Tur, Hoshen Mishpat, sec. 10, par. 3.

105 *Prisha* commentary on the Tur *ad loc.*

106 *Semah* on Hoshen Mishpat *ad loc.* agrees, par. 4.

107 Rashi, Jebamot 109b s.v. *ugemar halakha*.

108 Thus Haim Cohn's conclusion that Jewish law rejects the system of binding precedent is indeed supported by the example in Jebamot 109b. Haim Cohn, "Al Hatokim Atzmam Lidvar Halakha (the problem of precedent in Jewish law)," *Mishpat Vekalkala* 3 (1957), pp. 139ff. See note 41 above. Menahem Elon, on the other hand, in attempting to explicate the rejection of binding precedent, ignores the matter of "sticking like glue to a rule," cf. *op. cit.*, vol. 2, pp. 981ff.

The first example is the case of the squatter who resides on the vaçant property of another without his permission. There was an attempt to argue that the squatter must pay rent, on the basis of the decision handed down by R. Nahman in the case of one who built a mansion on land belonging to orphans. R. Nahman confiscated the mansion in lieu of rent.[109] But on closer examination of the alleged precedent it was found that Carmanians had been living on the land previously and had been paying a token sum as rent. Thus, although the land was not a real rental property, since it had generated only a nominal sum, it had the legal status of commercial property. The squatter was therefore obliged to pay for its use. The mansion was confiscated because the court instructed the trespasser to reach an agreement with the orphans and he failed to do so.

Another example is the case of one who planted trees on the property of another without the owner's permission. The Talmud deduced the opinion of Rab, which was not stated explicitly, from an earlier case,[110] in which Rab told the owner to have the planting assessed and pay its value to the planter. The owner refused, saying that he did not want the trees on his property to begin with. Rab again told him to have it assessed, assuring him he had the upper hand [i.e., that the owner would pay the equivalent of the value of the planter's investment or of the improvement of the property, whichever was lower]. The owner refused, insisting he wanted no planting on his property. Later, however, he built a fence around the plants and tended them. Rab told him, "By your act you declare that you do want the improvement. If so, have it assessed and pay the planter, but now, the planter has the upper hand."[111]

Clarification of factual differences is an integral part of determining the law. Imposition of an obligation to follow precedent would lull the judge into routinely deciding cases on the basis of authoritative recommendations.

R. Isaiah b. Israel Basan[112] does not mince words in excoriating blind reliance on precedent:

> All these judges who stick like glue to a particular responsum, without examining the roots of the matter, are not judges at all. The Ribash himself, in section 7, in replying to someone who wanted to rely on one of the Ribash's rulings in

---

109 bBaba Kama 21a, *Tosafot ad loc.* s.v. *veyahavei.*

110 bBaba Metzia 101a; *Shulhan Arukh*, Hoshen Mishpat, sec. 375, pars. 1, 3.

111 See Elon, *op. cit.*, vol. 2, pp. 960ff. Examples of distinguishing a precedent abound.

112 Eighteenth-century Italian scholar and teacher of R. Moses Haim Luzzatto.

> Barcelona, answered that it should not be used as a precedent because he might have been inclined to follow the Rashba of blessed memory on the matter, since Barcelona was under the latter's jurisdiction. And he may well have had his own reasons for deciding as he did in that particular case – perhaps due to an exigency. And as the Sages said,[113] one must not learn from a court action unless it is explicitly stated that the ruling should be implemented. Take heed of his words; they are as wise as their utterer and pleasing to men of understanding. Not everyone has the privilege of learning from the responsa of the great, since often their points are interdependent,[114] as it was stated, just as the fisherman gathers and keeps both the large and the small fish.[115] Know that the law varies with the circumstances of the case.[116]

I need add nothing to these persuasive arguments.

ii. Judicial Independence

The judge is not bound to follow decisions of other courts or even his own previous decisions. It is incumbent upon him to master and utilize the sources, and to learn from his teachers and the great halakhic authorities, but as a judge, he is independent – his decisions, and the responsibility for them, are his alone.

The Talmud reports:

> Rava said to R. Papa and to R. Huna b. R. Joshua: When one of my rulings is brought before you and you find something objectionable in it, do not tear it up before you come to me. If I have good reasons I will tell you; if not, I will revise my opinion. After my death do not tear it up, but do not rely on it either. Do not tear it up because if I were there perhaps I could have explained it to you; do not rely on it because a judge must rule solely on the basis of what his eyes behold.[117]

Commenting on this passage, the Early Authorities explain the phrase "do not rely on it" as follows: "Do not rely on it in another case, even if it is highly similar";[118] "Because even if we decide a case which is similar, the judge must rely only on what he sees";[119] "If another case is brought before him, even if it

113 bBaba Batra 130b.
114 Basically, obiter dicta.
115 bBaba Kama 41b.
116 Quoted in R. Isaac Lampronti's *Pahad Yitzhak* (Jerusalem: 1962), part 1, p. 325.
117 bBaba Batra 130b–131a.
118 *Nimukei Yosef* on Alfasi *ad loc.*
119 Ritba *ad loc.*, quoting his teacher R. Aaron Halevi.

is exactly like the first one, he may rule as he sees fit."[120] The *Yad Rama* concludes, "therefore the judges may decide on that case as they see fit."[121]

Nachmanides, in a responsum to the author of *Sefer Haterumot*[122] regarding a controversy between R. Yehudai Gaon and R. Isaac Alfasi, answered: "Though we may be in awe of the wisdom of the great master, there is no partiality in the law." R. Asher b. Jehiel puts it bluntly:

> If the judge does not agree with their opinion and he brings acceptable proof for his contemporaries, then Jephthah in his generation is like Samuel in his; there is no authority but the judge who is of these days, and he may contradict them. Because any point not explained in the Talmud edited by R. Ashi and Rabina can be contradicted and reconstructed, and even the opinions of the Geonim can be disputed.[123]

In a responsum addressing the question of whether an abusive husband, who starves his wife to the point where she hates her life, should be forced to grant her a divorce, R. Simon b. Tzemah Duran says:

> Though there are rulings of the Later Authorities, of blessed memory, to the effect that a man can never be compelled to divorce his wife, we are no reed-cutters [ignoramuses] either, and where it is a matter of reason, the judge must rule solely on the basis of what his eyes behold.[124]

Nor is the judge bound by his own former rulings on similar or even identical cases. As Rabbenu Gershom states: "A judge has to rely solely on what his eyes behold, that is, I decided according to what I thought fit at the time, at another time perhaps I would have regarded the matter differently."[125]

That it is conceivable for the same judge to rule differently on the same issue at different times is supported by the following example. The Talmud considers a case in which, after having left the court, the litigants lost the writ of judgment, and were unable to agree as to the substance of the ruling that had been handed down. The Talmud states that in such a case the judge is not

---

120 Ran *ad loc.*

121 *Yad Rama ad loc.*

122 R. Samuel b. R. Isaac Hasardi, *Sefer Hatrumot*, ch. 16, part 3; also in RCJ, vol. 2, p. 72.

123 *Piskei Harosh*, Sanhedrin, ch. 4, sec. 6. See Haim Cohn, *op. cit.* (note 108 above), p. 136; Elon, *op. cit.*, vol. 2., p. 985.

124 *Tashbetz*, part 2, sec. 8.

125 Rabbenu Gershom, Baba Batra 131a s.v. *sheein lo.*

to be trusted to say, "I ruled for this one and against the other." The Gemara asks: "Should he rejudge the matter?"[126] Rashi explains: "Either the same judge or another should decide in accordance with their claims."[127] The response in the Gemara is that this will not resolve cases where the decision is a matter of judicial discretion. Rashi explains:

> The outcome in such cases does not depend on reasoning but only on the mouth[128] of the judge. Take, for example, the case mentioned in bKetubot in which a man bequeathed his estate to Tobiah.[129] Two persons by this name appeared, both relatives and neighbors of the deceased. Such a case is decided by judicial discretion – the judge's intuition determines his choice. In effect he declares, "I think this man was a closer friend of the deceased," and awards him the estate. And since he has thus decided, even if he wishes to change his mind, he desists. Thus a new trial would not solve the problem, because this time he might decide in favor of the other party.[130]

Rabbenu Tam raises an objection: "If the outcome depends on the judge's discretion, as Rashi says, why doesn't he agree that rejudging the matter would serve its purpose, since presumably the judge would not change his line of reasoning?"[131] Apparently, Rashi thinks that a judge may change his attitude even with respect to a line of reasoning. As the Rashbam puts it, "The judge is to rule solely on the basis of what he sees with his own eyes. This applies to matters that depend on reasoning: he must decide only on the basis of what he sees."[132] We can conclude that judges are not bound by decisions of other courts on similar cases, or by precedents they themselves have set. Each case to be tried stands alone and must be decided on its own merits.

The following biblical exegesis is from Sifre:

> And I charged your judges at that time, saying: "Hear the arguments between your brethren, and judge righteously between a man and his brother."[133] And I told them: Be cautious in judging. If a case is brought before you two or three

---

126 bKidushin 74a.

127 Rashi *ad loc.* s.v. *venihader venidaynihu.*

128 In *Tosafot ad loc.* s.v. *beshuda dedayani*, Rashi's comment is quoted as follows: "everything according to what the judge beholds." This version is more likely.

129 bKetubot 88b.

130 Rashi on bKidushin 74a s.v. *beshuda dedayani.*

131 *Tosafot ad loc.* s.v. *shuda dedayani.*

132 Rashbam's commentary, Baba Batra 130b s.v. *sheein lo* (at the end of the page).

133 Deut. 1:16.

> times, do not say: This case had been brought before me and I know it inside out, but deliberate; be cautious in judging.[134, 135]

Elsewhere,[136] I used this principle, that every judge must try each case as if it were entirely new, to explain a query regarding a problem in bSanhedrin:

> There was a man they used to call "the mouse lying on money." Before he died, he mentioned several people to whom he owed money. After his death the creditors summoned his heirs before R. Ishmael b. R. Jose. He told them: There is a rule that states that a man is inclined to pretend to be poorer than he really is. (Rashi explains that the claim that one has debts is made to avoid the appearance of wealth and thereby ward off the evil eye.[137]) But this applies only in life, not after death. The heirs paid half, and were sued for the other half. The creditors appeared before R. Hiya, who ruled: Just as a man is likely to pretend to be poorer than he really is for his own sake, so he is likely to act for the sake of his children. They asked: Must we return the half already paid to us? (Rashi).[138] He replied: The Elder has already handed down his ruling.[139]

Now this report seems problematic. If a ruling had already been handed down by 'the Elder', that is, if the court had already ruled, given that such a ruling is final, it would not be necessary to litigate at all, as the rest of the amount would have been collectible on the strength of the first ruling. My understanding is that we are dealing with two creditors here, one of whom brought his suit before R. Ishmael b. R. Jose, won the case and recovered the money coming to him from the heirs, and a second creditor, who brought his case to R. Hiya. The latter did not take account of the precedent that had been set by R. Ishmael b. R. Jose's decision, and ruled differently on what was essentially the same case, rejecting the claim. On the basis of this second decision, the heirs sought to reverse the ruling of R. Ishmael b. R. Jose and recover the sum they had paid out. R. Hiya ruled as follows: "The Elder has already given his ruling" – his decision is final because this is not a case of an erroneous judgment, but "in a case like this, in which there is no explicit law,

134 See Mishnah Avot 1:1: "They [the men of the Great Assembly] said … be cautious in judging." Cf. R. Obadiah of Bertinoro's commentary *ad loc.* s.v. *hevu metunim badin.*

135 Sifre Deut., Dvarim, sec. 16.

136 See Z. Warhaftig, *Mimizrakh Noam Al Masekhet Sanhedrin*, p. 39.

137 Rashi on bSanhedrin 29b s.v. *adam asui* and *shelo lehashbia.*

138 Rashi *ad loc.* s.v. *nizol venihader.*

139 bSanhedrin 29b.

the judge must rule solely on the basis of what his eyes behold – and what he decides is final."[140]

For this reason I cannot agree with the conclusion of the *Orah Hamishpat*:

> If a case is brought before him which is identical with a case he has tried in the past ... the very same situation, and he knows that he gave his ruling for a certain reason ... but he cannot now recall that rationale ... he does not have to probe the case but may rely on his first decision. There is mishnaic support for this: If the anointed priest made an erroneous decision for himself unwittingly and acted accordingly, transgressing unwittingly, he must bring a bullock as a sin offering. If he made the wrong decision unwittingly but acted deliberately, or decided erroneously deliberately but acted unwittingly, he is exempt from a sin offering.[141] The Talmud objects: If he decided unwittingly and acted unwittingly, the Mishnah is redundant. Abbaye said: this matter is like the case of one who made a decision and forgot the reasons for it. When he acted in accordance with it, he said to himself: I am acting on the basis of my earlier ruling. The Mishnah teaches us that this case should not be compared to acting deliberately. It is clear that in relying on his earlier decision he should not be regarded as an intentional transgressor. If he should have waited until he remembered the reason for his earlier decision and did not do so but acted hastily, why is he regarded as an unwitting transgressor and not an intentional transgressor? We can infer that there is no duty to wait.[142]

But the *Orah Hamishpat*'s conclusion contradicts the passage from Sifre quoted above,[143] and the evidence it brings in support of this conclusion is very shaky, since the passages in question do not refer to cases that involve two actions, but rather, to an erroneous action that is repeated without having been corrected. For a decision holds even if the reasons for it are not given; therefore forgetting the reason, but handing down the decision nonetheless, is not a transgression.

In short, the judge must be independent and must rule afresh on every case that comes before him, as opposed to mechanically following precedent, whether his own or that of other judges. It follows that in Jewish law, precedent is intended only to guide future judges, not to bind them. The doctrine of binding precedent has no place in Jewish law.

---

140 Rabbenu Hananel *ad loc.*

141 Mishnah Horayot 6:2.

142 R.Y. Onalin, *Orah Mishpat*. R. Onalin was head of the rabbinical court of Tiktin and Schedlitz. Printed in 1908, this book was drawn to my attention by the late R. Efraim Zeev Garbuz.

143 See text above at note 135.

iii. The Judge's Duty to Hand Down a Ruling

The judge is obligated to render judgment. "Ye shall not be afraid of the face of any man," says the Torah.[144] R. Hanin tells us, "do not withhold your say because of any man."[145] Rashi explains: "The judge should not be afraid and withhold himself from handing down a decision."[146] That judges are required to render decisions is one of the 248 positive precepts.[147]

The duty to render a decision falls on the judge before whom the case is brought. Once the dispute is presented to him, he cannot relinquish the case.[148] The obligation to render judgment is a personal duty – the judge himself must decide, and the decision cannot be rendered by anyone else. R. Nahman said to Rava: "Did I not say to you that when I am sitting in judgment, you should not make any suggestions to me?"[149] From the verse "Thou shalt not follow a multitude to do evil,"[150] we learn that a judge must not follow the opinion of another judge, even one greater than himself, but every judge must decide the case according to his own understanding. This rule applies to all the great authorities, as the Tosefta states: "You should not say at the time of judgment: a slave may be content to be like his master (so too the individual should heed the opinion of the multitude[151]), but speak your mind."[152, 153]

The judge must be independent in reaching his decision. Over-reliance on precedent, excessive zeal in searching for pertinent references in the literature, and of course the need to rule in accordance with precedent – all these factors hinder independent judgment. For this reason, in capital cases, when the judgment is handed down, the decisions of the junior judges are given

144 Deut. 1:17.

145 bSanhedrin 6b. This is a play on the words of Prov. 10:5, *taguru* and *taagru* (bSanhedrin 7a).

146 Rashi on Sanhedrin 6b s.v. *deein lo ledayan.*

147 Rashi on Ketubot 106a s.v. *hai ase.* Similarly, following Onkelos, Rashi explains that the words of Exod. 18:22 (*veshaftu et haam bekhol et*) are in the imperative, see *ibid.* s.v. *veshaftu.*

148 bSanhedrin 6b. There is a special exception – suits suspected of being fraudulent, which judges may forgo (see Maimonides, *Code*, Laws Concerning the Sanhedrin 24:3–4); however, this exception is inapplicable to the present discussion.

149 bBaba Kama 96b.

150 Exod. 23:2.

151 *Minhat Bikurim* commentary on Tosefta, Sanhedrin 3:4 s.v. *dayo leeved.*

152 Tosefta, Sanhedrin 3:4.

153 See Warhaftig, *op. cit.*, pp. 50–52.

first.[154] The Jerusalem Talmud brings the following exegesis: "Do not follow a multitude... Rabbi says: the spelling of 'multitude' is the same as that of 'teacher' to indicate that you must not give your opinion after your teacher, but before him."[155] This is intended to keep the junior judge from being influenced by the senior, which would deter independence of thought and decision. To be sure, according to the Mishnah,[156] only capital cases begin with the junior judges, and in civil cases, the senior judge renders his decision first, but I have demonstrated elsewhere[157] that on the correct reading of this Mishnah, judgment in cases of civil law can commence with the decision of either the senior or the junior judge, as is deemed desirable.

Responsibility for the decision rests with the judge and is not shared with anyone else. The judge may, and indeed, ought, consult his teachers, and those of greater authority in his own town and throughout the country, but the decision and responsibility for the decision are his alone. The judge may, however, increase the size of the bench trying a given case, and thereby divide responsibility for it. "When a case was submitted to Rav Huna, he used to gather ten scholars in order that, as he put it, 'each might carry a chip of the beam.'"[158] But only the judges trying the case are responsible. The Rashbash concludes a responsum to a court as follows: "And what I have written is sufficient, though I have misgivings about this license, but the yoke is on you, who can see more where you are than I can hear from here, and I have saved my soul."[159] Even if it brings precedents for its decision, a court is not permitted to transfer responsibility, or even partial responsibility, to either the respondent or any authority he cites.

Resh Lakish explains the verse: "And judge righteously"[160] as "Consider rightly all the aspects of the case and afterwards decide."[161] The Tur and the

---

154 Mishnah Sanhedrin 4:2.

155 jSanhedrin 4:6.

156 Mishnah Sanhedrin, *loc. cit.*

157 Warhaftig, *op. cit.*, pp. 49–50.

158 bSanhedrin 7b; Rashi *ad loc.* s.v. *denimtein* explains that "we shall be burdened by a chip from the beam, that is, if we err, the offence will be divided among us all and my burden will be lightened". Compare the saying of R. Joshua b. Levi *ad loc.*: "Ten people who sit in judgment – the yoke [Rashi explains that this means 'the punishment for error'] is on all."

159 Rashbash, *Responsa*, #519.

160 Deut. 1:16.

161 bSanhedrin 7b.

Shulhan Arukh present this dictum as the law: "[The judge must] consider righteously in his heart and afterwards decide."[162, 163] The judge must arrive at and justify the decision himself – "in his heart" – and not rule as another court did.

We have seen that, for a variety of reasons, in Jewish law precedent serves as a guide, but is not binding. Against this background I will now briefly survey the relationship between Israel's Rabbinical High Court of Appeal and district rabbinical courts.

7. Precedent and the Relationship Between Israel's Rabbinical High Court of Appeal and District Rabbinical Courts

According to Jewish law, if they find its opinion unacceptable, courts are not bound to rule in accordance with any established authority, not even the decisions of a higher court, such as the Rabbinical High Court of Appeal. Secular Israeli courts, on the other hand, often accede to precedent: on more than one occasion, district courts have questioned the validity of decisions of the Supreme Court, but have considered themselves bound to follow it because of the doctrine of binding precedent.[164]

What happens when Israel's Rabbinical High Court of Appeal upholds an appeal against a ruling handed down by a district rabbinical court and returns the case to the district rabbinical court for retrial? There are cases where the court concedes the correctness of the decision of the appellate court, accepts it, and hears the case again. But there are also cases of courts that, convinced of the correctness of their decision as opposed to that of the appellate court, continue to firmly uphold their decisions and to maintain that the Rabbinical High Court of Appeal is mistaken. They do not question its authority or object on the grounds that courts are not subject to criticism, because an implicit condition of their jurisdiction is that litigants have the right to appeal their rulings; it is also one of the conditions they agree to in accepting their appointment as judges. There are thus no grounds for viewing this stance as an indication of contempt for the appellate court. These district courts argue

162 Tur, *Hoshen Mishpat*, sec. 17, par. 10.

163 *Shulhan Arukh* Hoshen Mishpat, sec. 17, par. 7.

164 Civil Case 778–876/50 A. B. v. M. B., District Court Reports, vol. 3, pp. 272–273; Application 1163/60 (Civil Case 1757/53) Shershevsky v. Shershevsky, District Court Reports 24, p. 183.

only that an appellate court which rejects a lower court decision should retry the case itself. They argue that retrial is not the responsibility of the lower court, which should not, and indeed, has no mandate to decide against its own convictions, if it has not been persuaded that its ruling was indeed wrong. In practice, the solution in such cases is usually for the appellate court to transfer the case to a different district court for retrial.[165] Thus, for example, the district rabbinical court in Haifa ruled that a husband be compelled to grant his wife a divorce.[166] Upon appeal, the appellate court overturned the ruling and instructed the district court to reopen the case in order to clarify the husband's claim that his wife had been involved with someone else even before their marriage. The presiding justice of the Haifa court refused to review the case, because in his opinion all the evidence had been carefully considered, and the ruling ought to stand as handed down. The case was transferred to the district rabbinical court in Tel Aviv-Jaffa.

One case, concerning custody of a child, raised more serious problems, when the lower court[167] refused to transfer it to a different court. In this case, the Rabbinical High Court of Appeal intervened – before the decision of the district court had been handed down – when one of the parties appealed a decision of the district court concerning the deposing of witnesses abroad. While the Rabbinical High Court of Appeal rejected the appeal, it nevertheless ruled that the court should transfer the case to a different court. The district court argued that a case under adjudication should not be transferred, and that the court hearing a case has a duty to pass judgment on it. The matter reached the Supreme Court,[168] which instructed the district rabbinical court in Jerusalem to heed the Rabbinical High Court of Appeal. The Supreme Court did not address the halakhic approach to the question of the extent to which decisions of one court are binding on other courts. It simply ruled that a

165 E.g., Files 37/3692, 39/1401, Haifa District Rabbinical Court.

166 File 34/202, Haifa District Rabbinical Court.

167 File 1969/2312, Jerusalem District Rabbinical Court.

168 High Court of Justice 35/72, Kalir v. Jerusalem District Rabbinical Court and Kalir, P.D. 26(1), p. 757.

decision of the Rabbinical High Court of Appeal is binding even if based on a mistake of law, as long as it does not exceed its authority.[169]

Such exceptional cases highlight the judicial independence of the rabbinical courts.

169 *Ibid.* p. 759, "gimel-dalet." It is noteworthy that at the end of its ruling the Supreme Court states:

> There is no need to point out that given the existing complex hierarchy of legal institutions, it is incumbent on the court of the first instance to heed the decisions of the appellate division. The debate which the first respondent (the Jerusalem District Rabbinical Court) entered into with the appellate institution above it is unseemly, and it ought not be the case that such a matter should require a petition to this court to compel the first respondent to comply with the ruling of its appellate court (pp. 759–760).

The Supreme Court ordered the judges of the Jerusalem District Rabbinical court to pay the court costs of the appeal to the Supreme Court.

CHAPTER 2

THE JUDGE–AGENT ANALOGY IN THE TALMUD

Hanina Ben-Menahem

1. Introduction

A number of passages in the Talmud[1] intimate that the institution of agency is related to that of the court of law. The Talmud equates judges with agents in a number of places, and derives laws pertaining to courts from the laws of agency. Conversely, the latter are sometimes derived from the former. However, the analogy between courts and agents is not universally accepted by the Sages. Nevertheless, if we are to gain an accurate understanding of the Sages' jurisprudential outlook, it is important to take note even of those concepts, analogies and hypotheses that are ultimately rejected or fail to gain unanimous recognition. Halakhic debate can teach us as much about the theoretical background of talmudic law as can established legal decisions.

1 These passages will be discussed below.

This study attempts to clarify the conceptual aspects of this analogy, its scope and limits, and its influence on the status of judges in Jewish law.

Formally, the analogy between judges and agents is based on the fact that both act on behalf of others rather than for themselves; agents act for their principals, judges for the litigants.

However, as I discuss below, the analogy appears to break down when the key element of consent is taken into account. One cannot act as an agent against the wishes of the principal, nor can one be appointed by a principal to serve as his agent against one's wishes. The consent of both agent and principal in the establishment and continuation of the relationship between them is inherent in the institution of agency.

This element does not seem to obtain with regard to agency mandated by law, i.e., instances when the law declares that specified actions of A, the agent, are to be viewed as having been performed by B, the principal. This type of agency, which is constituted by law without regard to the wishes of the principal, is sometimes called "constructive agency."[2] Nevertheless, constructive agency and garden-variety agency have in common the fact that the act performed by the agent is considered to be an act of the principal. Thus, for example, a widow who sells assets from the estate of her late husband in order to realize her marriage contract acts as an agent of the heirs, though they did not appoint her as such and might even oppose her action.[3] The law regards her as the agent of her late husband's heirs. They are, indeed, the beneficiaries of her husband's estate, and obligated to pay her share, but in order to protect the widow and prevent the heirs from withholding, or delaying, assets or monies they are legally obliged to settle on her, she can sell part of the estate herself to realize her marriage contract. By regarding her as acting on behalf of the heirs, the law recognizes the widow's right to resort to self-help to realize her claim.

In both agency and constructive agency, the authority of the agent rests on the explicit or constructive agreement of the principal, and in both cases the action of the agent is attributed to the principal. The agent has no independent authority whatever. This is reflected in the fact that there is no impediment to the principal himself performing the action he has delegated to his agent. However,

2 See A. Barak, *Hok Hashlihut 5725–1965*, ed. G. Tedeschi, (Jerusalem: 1975), p. 5.

3 Mishnah Ketubot 11:2–4.

with regard to constructive agency, the principal cannot prevent the agent from performing the task in question if the principal himself does not perform it.

2. Consent of the Principal and Judicial Independence

The first obstacle to accepting the Talmud's analogy between judge and agent is conceptual. If a judge is comparable to an agent – an agent of the litigants, it would seem – then the judge's authority to act derives from the consent of the litigants. Even if this consent is only constructive, nevertheless, this still means that the judge has no authority beyond that bestowed upon him by the litigants. The judge thus lacks autonomy, and is not independent of the litigants. These implications of the judge–agent analogy are incompatible with the usual conception of judicial authority, under which judicial authority is not contingent upon any authorization by the litigants.

There are four directions that can be explored in seeking a resolution of this incompatibility:

1) The first possible approach is to deny that there is a full legal analogy between judge and agent and to interpret the Talmud's remarks as intended metaphorically rather than literally; as descriptive rather than prescriptive. This approach is adopted by R. Moses Sofer, who argues that the dictum "we are acting as their agents,"[4] establishing the jurisdiction of the Babylonian Sages, is not meant in the legal sense because "the principals are long dead, while the agents are as yet unborn, and no such agency is known to us."[5] This strategy counsels overlooking the text's apparent meaning because it is problematic; it thus undermines the reader's communicative rapport with the text, and is, to that extent, an unattractive solution.[6]

To be sure, the model of agency is indeed cited by the Talmud in a variety of contexts, some of which seem remote from its ordinary definition in the

4 bGitin 88b, bBaba Kama 84b.

5 Hatam Sofer, *Responsa*, Orah Haim #84. For another example, see #201. Similarly, the Hida, (R. Haim Joseph David Azoulay) referring to an attempted deduction of a principal–agent relationship between Jews and Gentiles from the Talmud at bAvoda Zara 53b, asserts: "What is said – that they serve as agents – is a descriptive metaphor, not meant in the legal sense" (*Birkei Yosef*, Even Haezer #16 s.v. *vekhol*). I would like to thank Dr. Daniel Sinclair for drawing the latter reference to my attention.

6 For another example of this type of difficulty, though in the biblical context and not the talmudic, see the responsum of R. Samuel b. Hofni relating to the diviner of Ein Dor in B.M. Lewin's *Otzar Hageonim* on Hagiga, pp. 2–3.

legal context. Indeed, the concept of agency is used by the Talmud to explicate a diverse group of rather obscure legal institutions and relationships.[7] The Talmud seeks to illuminate these complex relationships, to explain these unfamiliar concepts, by citing a concept which is established and familiar – that of agency – even though its introduction raises new difficulties.

This problem is related to the wider question of the explication of legal institutions. It is difficult to know how much importance the Talmud attributes to such explications. The literature reveals that commentators on the Talmud indeed deliberate on the question of whether the general model of agency in its various contexts is merely metaphorical or has actual legal application.[8]

2) Conversely, it could be inferred from the talmudic analogy that judicial authority does in fact emanate from the litigants' consent and that, contrary to the received view, the judge indeed lacks any independent status. This is proposed by Asheri on the authority of the Tosafist R. Isaac: "the judge's authority does not exceed that of any other agent."[9] This opinion is adopted by Albeck, for reasons of his own.[10]

3) The third possible solution is a compromise between the first two suggestions. The analogy is accepted only partially. Judges are not construed as agents in all contexts, but only within a restricted range of instances where the two concepts are explicitly associated. However, this solution is also less than ideal, as the following case illustrates.

In a discussion of fraud (*onaa*), the Talmud compares the status of a widow to that of an agent.[11] By law, fraud is not recognized when an owner of immovable property sells it himself; all such sales are thus final. However, if a widow sells property belonging to her husband's estate, such a sale is void if the price differs even slightly from the market value; that is, in this case the ordinary laws of fraud are superseded. The same is true in the case of such a transaction when carried out by an agent.

In the wake of the Talmud's conclusion that the status of an agent is identical to that of a widow, R. Hai Gaon draws on the analogy to infer that: "the

7 See bAvoda Zara 53b, bJebamot 10b, bGitin 21a.

8 See, for example, R. Haim Joseph David Azoulay, *loc. cit.*; the *Porat Yosef* commentary on bKidushin 23b; and *Avnei Nezer*, Even Haezer #223:12–13.

9 Asheri on Ketubot 100a; cf. *Tosafot ad loc.* s.v. *Rava*.

10 Cf. S. Albeck, *Batei Hadin Bimei Hatalmud* (Ramat Gan: 1980).

11 bKetubot 100a. I use the term "fraud" to refer to *onaa*, the act of harming another by selling one's personal property for more than its actual value. This is elaborated on below, sec. 3. ii.

agent must effect the sale in the presence of three ... just as is required of the widow."[12] R. Isaiah Di Trani comments:

> I disagree. The Talmud only likens the laws of agent and widow with regard to fraud, but not with regard to the requirement of a quorum of three. Since the agent is employed by the principal, he is permitted to sell by himself. However, the widow sells against the wishes of the heirs, so the sale must be performed in the presence of three. The agent is not like the widow in this respect.[13]

Apparently, R. Isaiah Di Trani subscribes to the idea of partial analogy. But his position, grounded on the distinction between a sale carried out against the will of the principal – namely, the widow – and one carried out with the principal's blessing, is open to criticism. For this argument could apply equally well to fraud: were this the case, deviation from market value would void the widow's transaction, but not that of the agent. However, the Talmud does not accept the distinction, and, considering the similarities in the two cases to outweigh the differences, maintains that agents are comparable to widows with respect to the law of fraud. If so, the opinion of R. Hai Gaon is preferable from a logical point of view.

4) Another possible strategy is to say that the judge is indeed an agent, but not of the litigants. Rather, he is an agent of the public or the Almighty, as, for example, in the case of priests, of whom it is said: "these priests are the agents of the Almighty."[14] The implications of construing judges as agents of the public or of the Almighty, respectively, will be explained below.

Before turning to the talmudic passages which advance the agent–judge analogy, I want to point out that these passages are all from the Babylonian Talmud and lack parallel versions in the Jerusalem Talmud. It remains to be determined whether there is any connection between the view that judges are comparable to agents and the Babylonian Talmud's position on the question of the source of judicial authority, according to which the authority of the judges in Babylonia derived from their serving as agents of the judges in the land of Israel.[15]

---

12 R. Hai Gaon, *Sefer Hamikah Vehamimkar*, 6. Sale by a widow is permitted only in the presence of three others, and not if she is alone. Cf. bBaba Metzia 32a.

13 *Tosafot Rid*, Ketubot, *ad loc.*

14 bYoma 19a, bKidushin 23b, bNedarim 35b.

15 bGitin 88b, bBaba Kama 84a. The model of agency is used even to elucidate relationships among the Babylonian judges themselves, see bHulin 18a: "we are acting as their agents." Cf. Mishnah Yoma 1:5, bYoma 19a.

3. Talmudic Passages Suggesting the Judge–Agent Analogy

i. Judicial Discretion

"With regard to two documents bearing the same date ... Samuel says it is left to the discretion of the judges (*shuda dedayani*)."[16] That is, the judges decide which document is to be endorsed. The Talmud objects, quoting a *baraita* which states, "In a case of two documents bearing the same date, the assets are divided."[17] The Talmud concludes that the Tannaitic debate on the issue is unresolved, and that Samuel rules according to the Tanna who advocates judicial discretion (*shuda*). According to the Gemara, this Tanna is the one whose opinion is cited at the end of the following *baraita*:

> If one man said to another, Take this money (*mane*) to so-and-so, and he went and sought him but could not find him, then the coin must be returned to the sender. If the sender has died – R. Nathan and R. Jacob say: to the sender's heirs; others say: to the heirs of the recipient ... the Sages say: it is to be divided; and here it was said that the agent should do whatever he wishes.[18]

Thus the Gemara likens Samuel's law regarding judicial discretion to that in the *baraita*, which deals with the discretion exercised by agents.[19] This fundamental rule governing the decision-making process to be used by courts is

---

16 bKetubot 94a. For the etymology of "*shuda dedayani*," see my *Judicial Deviation in Talmudic Law* (Chur, Switzerland: 1991), p. 82. For a general discussion of this concept, see *ibid*., pp. 150ff.

17 *Ibid*., 94b. The Jerusalem Talmud rejects the opinion of Samuel on the basis of a Mishnah which contradicts it. See Alfasi on Ketubot 85b.

18 The complete *baraita* appears in bGitin 14b with some textual variants, e.g., "third party" for "agent." Cf. *Dikdukei Soferim ad loc*. "Here it was said" apparently refers to Babylonia; see R.N. Epstein, *Mevoot Lesifrut Hatanaim*, (Jerusalem: 1957), pp. 245–6; M. Higger, *Otzar Habaraitot*, vol. 8, p. 151.

19 According to bGitin 14b–15a, the Sages who maintain that the money should be divided do so because of doubt as to a point of law. (There are two accounts of the question over which this doubt arises. On one account, the question is whether the agent serves as an agent of acceptance for the recipient as well as an agent of the sender; on the other, it is whether the law is decided according to R. Eliezer.) According to the Sage who states that the agent should have discretion, this solution is to be preferred when such legal doubts arise.

According to bKetubot 94a, however, the dispute between Rab and Samuel is over a factual matter (cf. Tosefta Baba Batra 13:8). The solution, according to the

inferred from the laws of agency, and the Gemara does not question the validity of the judge–agent analogy.[20]

The *baraita* of the agent basically describes a case in which the principal fails to provide the agent with adequate instructions. The instructions supplied are deficient in that they do not provide directions for circumstances in which the original mission cannot be fulfilled. The Tannaitic opinion that "the agent should do whatever he wishes" suggests that the laconic instructions of the principal are to be regarded as authorizing the agent to use his discretion if the primary instructions cannot be fulfilled. If the instructions are intentionally laconic, then it is reasonable to assume that the agent is, from the outset, granted discretionary authority to decide his course of action. Even if the instructions were given without anticipating the problems that later ensued, impeding their fulfillment, they imply that the agent is indeed granted such implicit authority.

The model of implicit authorization requires assessing what the principal's wishes can be presumed to have been, giving due consideration to the fact that, had he so desired, the principal could have denied the agent this discretionary authority. Indeed, it is clear that the principal has the power to deny the agent such authority not only by making an explicit statement to this effect, but also by providing clear and comprehensive instructions on how he should proceed in the event that the primary instructions become impossible to execute.

What type of case would be analogous in the judicial context? A case where instructions issued by an individual are insufficiently clear and hence must be interpreted by a court. The Talmud records such a case, that of one who says, "Give a *mane* to Reuben," and then dies, whereupon two men by the name of Reuben come forward, and there is no way to determine which of

---

Tanna whose opinion is introduced by the phrase "And here it was said," is to grant the agent discretion.

The two passages thus interpret the *baraita* differently, though it is clear that the discussion in bGitin reveals awareness of the interpretation offered in bKetubot.

20 Nahmanides' commentary on Baba Batra 34b s.v. *umahi shena*, suggests the following interpretation of the *baraita*: "In Ketubot too it is said, the third person [agent] does as he wishes, that is, the court awards it to the more virtuous one." He thus seems to maintain, perhaps due to the difficulty pointed out above, that the court, and not the agent, has the authority to decide.

them is the intended beneficiary.[21] On the judge–agent analogy model, the court is the testator's agent. The laconic instruction constitutes an implicit authorization for the court to decide as it sees fit, to use its discretion. This authority to determine the testator's wish and to effectuate his instructions is bestowed upon the court on the premise that it is at the testator's behest that the court accepts the commission of interpreting his wishes and conferring effective, binding validity on them. The analogy will fail, however, if such implicit authorization cannot be assumed. This might occur in a case of a disputed land claim, where two parties both contend that the land in question is their inheritance, for in this type of case there is no vague or laconic instruction that can serve as an implicit authorization.[22]

Indeed, with one exception, the cases in the Talmud where the situation giving rise to the application of the principle of judicial discretion is described all involve determination of the wishes of an individual who has expressed himself laconically or unclearly, inviting the inference that the court has been implicitly authorized to intervene and carry out these wishes at

---

21 bKetubot 85b.

22 A similar distinction can be extrapolated from Alfasi's analysis of bKetubot 85b, in which he states that the principle of judicial discretion applies only to cases involving documents such as deeds which bestow mutually exclusive rights. For only then can it be argued, "if it belongs to A, it does not belong to B, and vice versa." But in cases of documents, such as loan agreements, which attest to non-mutually-exclusive interests, both parties could theoretically have legitimate interests; in such cases, the policy governing the disposition of the assets is not judicial discretion, but rather, division of the assets between the claimants.

This distinction is in harmony with my argument that judicial discretion is only applicable if an implicit authorization can be presumed. Given that an asset that has been sold or given to one person could not also have been sold or given to the other, the existence of what appear to be conflicting yet equally valid documents compels us to assume that the court has been implicitly authorized to decide which is to be accepted, not unlike the case of "Give a *mane* to Reuben," when it is not specified to whom "Reuben" refers.

On the other hand, different obligations to different people do not necessarily contradict each other, and in this type of case, therefore, there is no reason to assume that a court has been implicitly authorized to decide according to its discretion. The fact that a creditor may have committed himself financially far beyond his ability to meet these obligations is an external factor unrelated to the actual generation of these obligations; hence, it does not require us to assume the existence of an implicit authorization.

its discretion.[23] The only case invoking the principle of judicial discretion in which such an authorization cannot be assumed, is the following: "One says, it belonged to my ancestors, and the other says, it belonged to my ancestors."[24] In a case such as this, in which both parties claim ownership on the basis of inheritance, but neither can prove his claim conclusively, the Talmud says: "The stronger has the upper hand." This declaration is then questioned: "Why should the ruling in this case differ from that in which two deeds relating to the same property bearing the same date are presented in court ... regarding which Samuel says 'judicial discretion'?" The Talmud replies that the latter situation is different because there the relevant claims can, in principle, be verified.

Thus the conclusion of the Talmud is in agreement with that of the analysis suggested above, namely, that the principle of judicial discretion cannot be employed in cases which do not involve the interpretation of laconic instructions.[25]

ii. Agents as Judges

With regard to fraudulent land transactions, the Talmud distinguishes between transactions carried out by the owner of the property and those carried out by a third party acting for the owner.

---

23 bKetubot 84b, 94a–b, bKidushin 74a, bGitin 14b–15a, bShevuot 30a–b. The dispute in bBaba Batra 62b which calls for judicial discretion is a later, Saboraic, interpolation; cf. *Nimukei Yosef ad loc.*

24 bBaba Batra 34b–35a.

25 It could be argued that the Talmud defines the principle that in the absence of vague or laconic instructions the court has no authority to exercise discretion. As we said, the court's authority to decide is based on the assumption of implicit authorization, which originates from deficient instructions. On the other hand, it could be argued that were this true, the Talmud's question could have been answered more directly; however, it may well be that the Talmud prefers to answer in a way that not only makes it clear that an instruction is needed, but also indicates exactly what type of instruction is required.

Furthermore, the argument that judicial discretion cannot be applied in a case of "One says: it belonged to my ancestors," is intended to prevent subsequent controversy over the court's decision should additional witnesses materialize later on (cf. Rashbam *ad loc.* s.v. *hatam*). There is no concern about such an eventuality in the other cases, where judicial discretion is sanctioned, because in these cases the judges are only assessing the subjective intentions of the principal, a matter about which witnesses cannot testify. Consequently, the court's judgement cannot be impugned.

In the former, fraud is not recognized and the sale is valid.[26] In the latter, the validity of the sale depends on the identity of the third party. If a widow, authorized by the Sages to sell land belonging to her husband's heirs to realize her marriage contract, sells property worth a *mane* and a *dinar* for only a *mane*, the sale is void.[27] On the other hand, a sale by the court is only void if it deviates from the market value by a sixth or more.[28]

On the basis of these laws, the Talmud investigates the validity of sales by an agent. "It was asked: To whom is an agent comparable? Rava in the name of R. Nahman: an agent is like a judge. R. Samuel b. Bisna in the name of R. Nahman: an agent is like a widow."[29] The Talmud goes on to explain the reasons for the disagreement between Rava and R. Samuel b. Bisna. "Rava says, the agent is like a judge, because neither the judge nor the agent has any personal interest in the transaction." Obviously, this alone does not suffice to establish the judge–agent analogy. An implicit premise must be assumed here: that judges and widows are agents, hence their relevance in deducing the status of other agents, who are not judges or widows. However, each has distinctive characteristics, and therefore, the laws pertaining to them differ.

The question is which of the two best exemplifies agency. Rava thinks that the widow's personal interest in selling the land exceeds any agent-like feature she embodies, which explains why her case is an exception. In rejoinder to the argument that the law of agents should be deduced from that of widows, since "the widow looks out for her own interests," the Talmud informs us that judges and agents do not seek to further their own interests.

R. Samuel b. Bisna, on the other hand, maintains that as a public institution, the court is distinguishable from the widow and the agent, who are private individuals. He too contends that judges are agents, but argues that the public aspect of their role adds another element to their status. As R. Menahem Hameiri points out: "Though judges are like agents, nevertheless, we give them more power and authority than we do other agents. Hence their transactions are not void unless they deviate by more than a sixth."[30]

26 bKetubot 99b.

27 *Ibid.* 98b.

28 *Ibid.* 99b.

29 *Ibid.* 99b–100a. My account is line with that of those commentators, among them Rashi and R. Isaac the Tosafist, who interpret the talmudic query as addressing the status of agents in general, and not just that of bailiffs (court agents).

30 This shows that even the Sages who oppose R. Simon b. Gamaliel in the Mishnah accept the principle of "if so, what good is the court's authority?"

iii. Agent as Witness and Witness as Judge

The Talmud reports a debate about the composition of the three-member court which must be present when an agent gives a woman her writ of divorce. R. Johanan says that the presence of two, in addition to the agent, suffices; R. Hanina requires three. The Gemara tries to identify the reason for the disagreement. One suggestion is that:

> The dispute is about whether an agent can act as a witness, and a witness can act as a judge. He who says, before two, asserts that an agent may act as a witness and a witness as a judge; while he who says, in the presence of three, agrees that an agent may act as a witness, but holds that a witness may not act as a judge.[31, 32]

On this standard version of the explanation for the dispute, the argument is over the question of whether a judge can be compared to an agent, and if so, whether it can be inferred that since an agent can serve as a witness, a witness may act as a judge. As Rashi explains: "if we say that in such circumstances an agent may become a witness, it could also be said that a witness may become a judge."[33] In other words, if the relationship between principal and agent does not disqualify the latter from testifying in favor of his principal, then the relationship of judge and litigant ought not disqualify the judge from testifying for one of the litigants.[34] The argument is premised on the assumption that the

---

and their argument with R. Simon b. Gamaliel has to do with the scope of this principle. R. Simon b. Gamaliel contends that on the strength of this principle, a deviation of up to 50 percent in a transaction conducted by a court is acceptable (bKetubot 100b) while the Sages limit permissible deviation to a sixth.

31 bGitin 5b.

32 Cf. the following external sources: Plato, "If one of the judges is called upon to testify, he may testify but not vote in that case (*Laws* 937a); Harrison, *The Laws of Athens* (Oxford:1971) p.137; A. Gellius, *The Attic Nights,* vol. 3, p. 27. It is interesting to note that Gellius addresses four questions of interest to the Sages: the witness who serves as judge; whether a judge should mediate or adjudicate; whether a judge can take the counsel's part; and whether the esteem in which a court is held is a consideration in its decision-making.

33 Rashi, bGitin 5b s.v. *beshaliah neesa ed veed neesa dayan.*

34 It is clear that the Talmud attempts to connect the problem of whether an agent may serve as a witness to that of whether a witness can serve as a judge, as Rashi states. It cannot be maintained that according to the Gemara, R. Johanan and R. Hanina differ only on the question of whether a witness can serve as a judge. For were this the case, the Gemara would have said, "the dispute is about whether a witness can serve as a judge," rather than, "the dispute is about whether an agent can serve as a witness, and a witness a judge."

concepts of judge and agent are identical, since otherwise the inference from the law governing the one to that governing the other would be invalid.

R. Johanan is of the opinion that if the law allows an agent to be a witness, then we must also mandate that a witness can serve as a judge; hence, the agent who serves as a witness can also, along with two others, serve as a judge. It thus suffices for the writ of divorce to be delivered in the presence of only two other judges.

The Talmud's rejection of this explanation lends credence to this account of the dispute. The Talmud says that with regard to Rabbinic law it is unanimously accepted that a witness can serve as a judge, and it simply cannot be maintained that there is any controversy whatsoever on this point.

From our point of view, the version of this passage found in *Halakhot Gedolot* is particularly interesting. "The one who says, before two, is of the opinion that an agent can act as a witness and a witness as a judge; and the one who says, before three, is of the opinion that an agent may not serve as a witness and a witness may not serve as a judge."[35] This version suggests that the dispute between R. Johanan and R. Hanina is not over the validity of the judge–agent analogy. On the contrary, it maintains that they are in agreement that the analogy is feasible.[36] Rather, on this version, the dispute is over the specifics of the law governing agents, from which that governing judges is derived.

iv. The Presumption that Agents Discharge their Responsibilities

The Talmud records a debate between R. Nahman and R. Sheshet concerning the scope of the presumption that an agent discharges his responsibility.[37] R. Nahman contends that it applies only to Rabbinic law, while R. Sheshet thinks that it applies to biblical law as well. The Talmud explains the view of R. Sheshet as follows:

> R. Sheshet says: On what grounds do I say this? For we have learned in a Mishnah: "The new harvest is permitted immediately following the Omer sacrifice. Those who are far away are permitted to consume the new harvest from noon onwards." The law of the new harvest is a biblical law, yet the Mishnah states that those who are far away are permitted to consume the new harvest from

35 *Halakhot Gedolot*, Gitin, Hildesheimer ed., p. 160.

36 The analogy between the passage which discusses "the agent who acts as a witness" and that which speaks of "the agent who acts as a judge" is also reflected in the language used. Cf. the *baraitot* in bKidushin 43a and bRosh Hashana 25b.

37 bEruvin 31b–32a.

noon onwards. Is this not because of the presumption that an agent discharges his responsibility?[38]

R. Sheshet infers the scope of the law of agents from the law of the Omer sacrifice, for which the courts are responsible. This inference is possible only if the judge–agent analogy is recognized.

R. Nahman rejects the analogy: "The Mishnah says so for a reason – it is known that courts do not procrastinate." R. Nahman thinks that a distinction should be made between judge and agent: a court may be relied upon to discharge its responsibilities, but with regard to an individual agent, prompt discharge of his responsibilities should not be presumed in matters of biblical law.[39]

The Early Authorities ask an interesting question about the opinion of R. Sheshet: "How can the Mishnah be used as a proof, since in the case discussed there, the priests [who compose the court] are not only acting as agents but also are obliged to perform the Temple service?"[40] The agent–principal relationship is voluntary, being constituted by the free choice and consent of (at least) the agent, but the priests lack this freedom with regard to Temple responsibilities, which they are duty-bound to perform on behalf of the people. Therefore the law governing them cannot be used as a basis for the law governing agents who carry out their responsibilities voluntarily.

To resolve the difficulty, a distinction is made between the duty to perform the Temple sacrifices, and the time and manner of their performance. From the fact that this duty, which can be carried out any time during the day, is discharged, we indeed cannot infer that there is a presumption that an agent discharges his responsibility. However, the fact that the priests hasten to perform their duty, in fact, to perform it before noon, so that everyone can partake of the new crop, constitutes a voluntary element in the priests' actions which justifies the analogy with voluntary agency.[41]

---

38 *Ibid.* 32a.

39 R. Nahman's stance here resembles that of R. Samuel b. Bisna in the name of R. Nahman, cited above: "An agent is like a widow and not like a judge, because the court is composed of many." Cf. above near note 29ff.

40 R. Yomtov b. Abraham of Seville (Ritba), *ad loc.* Cf. *Tosafot* on Eruvin 32a s.v. *harehokim*; and R. Solomon b. Adret *ad loc.*

41 This might be the reason why Rashi (*ad loc.* s.v. *harehokim*) states that "priests are our agents who discharge their responsibilities," even though the conclusion of bKidushin 24b is that priests are agents of the Almighty. But, as we said, the inference is not from the fact that they discharge their Temple duties, a responsibility with which they are charged by the Almighty, but rather, from their haste to do so before noon, in which they act as agents of the people.

4. Conceptual Difficulties with the Analogy

Having examined a number of passages which suggest the judge–agent analogy, we now turn to some additional conceptual difficulties which must be overcome if the analogy is to stand. The model which construes the judge as agent of the litigants is feasible only if the judge is engaged on the initiative of the litigants. If this is the case, it can be argued that the judge's authority indeed rests on the authorization granted him by the litigants. It is difficult to apply this model, however, if the court has initiated the action, in which case there is no room for the premise that the litigants themselves grant this authorization.

Logically speaking, it is possible to distinguish two types of courts: those which are voluntary judicial institutions, which take upon themselves the responsibility of adjudicating the cases brought before them,[42] and the litigants commit themselves to accepting the courts' rulings as binding; and those courts which function as the judicial arm of a society or government. The establishment of these courts does not rest on the endorsement of the individuals whose cases they try.

This distinction is useful in facilitating our understanding of the various talmudic sources dealing with this subject and in resolving many of the problems that have been noted. On the conceptual level at least, it appears that the distinction was recognized by talmudic law. We find some sources concerned with the regulation of procedures followed by voluntary courts, while others pertain to courts that function as judicial arms of society, and it is clear that no inferences are drawn from laws pertaining to one type to those governing the other.[43] Consider the following passage:

---

42 An example of this in Israeli law is the Law of Arbitration (5728–1968) which confers the force of a legal ruling on any decision resulting from a voluntary process before one who consents to serve as a judicial organ, provided that the decision is not contrary to law. Note, however, that such a decision cannot be enforced without the permission of the district court, that is, the process of arbitration is subject to state judicial review. Moreover, if a party to such voluntary arbitration claims the arbitrator has perpetrated a gross miscarriage of justice, it has the right of appeal to a state court.

43 A separate question, which I do not cover here, is that of the historical and sociological factors which facilitated the conceptual development of the two types of courts. It is reasonable to conjecture that the absence of public courts was instrumental in the growth of voluntary courts. Cf. A. Gulak, *Yesodei Hamishpat Haivri* (Berlin: 1922), vol. 4, pp. 30ff.

> Two come to you for adjudication, one gentle, the other violent. Before hearing them, or even after hearing them, if you are still undecided, you are permitted to tell them: "I do not want to be involved with you." Lest you find against the violent litigant, and he persecutes you.[44]

Obviously, this passage refers, not to a court functioning as the legal organ of the state, but to a voluntary court, which, as noted above, requires the assent of the judge. In a similar vein, there is a Mishnah which states: "This one chooses one, and the other chooses one, and the two choose a third one together."[45] The Talmud explains: "What is meant by the statement that each chooses a judge? [Here Rashi notes, "it implies that this is the norm."] The meaning is: If each has chosen a judge, then the third judge must be chosen jointly."[46] That is, the Mishnah indicates a procedure for the consensual adjudication of monetary disputes.[47] As a final illustration, consider acts which by law are legally valid only if performed in court. The Talmud states that some such acts may be performed before a voluntary court. For example, "R. Judah said: Three who entered to visit an ailing man, they may draw up documents or convene as a court if they wish."[48]

All these examples do not, of course, warrant the conclusion that the other, non-voluntary type of courts was unknown to the Talmud.

Let me summarize. The judge–agent analogy is only relevant in the context of the first type of court. Judges who officiate in such courts can be characterized as follows: they are not duty-bound to adjudicate, and their authority rests on the consent of the litigants. The analogy is much more difficult to uphold if the judges in question preside over courts of the second type, that is, if they are duty-bound to adjudicate and their authority does not rest on the litigants' consent.

Note that the attempt to conceive the litigant–judge relationship as a contractual nexus[49] is problematic even within the limited context of voluntary adjudication. Talmudic law does not permit a judge to receive remuneration for his services,[50] but only compensation for lost time, that is, the sum the

---

44 bSanhedrin 6b.

45 Mishnah Sanhedrin 3:1.

46 bSanhedrin 23a.

47 This point is emphasized by the phrasing of the passage, see Rashi and *Tosafot ad loc.* s.v. *mai.*

48 bBaba Batra 113b; *Tosafot* and Rabbenu Gershom interpret the passage differently.

49 Albeck, *op. cit.* (note 10), pp. 52–57.

50 Mishnah Bekhorot 4:6, bKetubot 105a, bKidushin 58b, jSanhedrin 1:1.

judge would have earned by pursuing his usual occupation. If the usual occupation of the judge is arduous physical labor, which he interrupts in order to adjudicate, then the litigants do not reimburse him for the entire loss of his wage, because sitting in judgment is less strenuous than his usual occupation, and the judge need not be compensated for this leisure.[51] The principle is that the only permissible payment is a sum which leaves the judge in the same financial situation he would have been in had he not engaged in adjudication. The judge must not derive any financial advantage from his agreement to sit in judgment.

Essentially, contractual relationships are relationships from which both parties intend to benefit; an example would be a craftsman who receives a wage for services rendered. This does not negate the possibility of one's entering into a contractual relationship which in the end – or even a priori – brings him no benefit. I am not claiming that the element of consideration is a sine qua non in Jewish law. But it is difficult to see how a relationship in which one side can never advance its interests can be considered a contractual relationship. If the judge is forbidden to derive any benefit from his agreement to carry out this function, in what sense can his relationship with the litigant be construed as contractual? The theoretical dividend from describing the judge–litigant relationship as contractual is at best minimal. Furthermore, the prohibition against accepting wages is based on the idea that adjudicating is a religious duty incumbent on all those who are qualified, and it is prohibited to accept payment for the performance of a religious duty.[52] In other words, the judge does not take upon himself a new obligation by his assent to adjudicate, because the obligation exists in any case.[53]

51 bBekhorot 29b; also see the Early Authorities *ad loc.*

52 bBekhorot, *ibid.* Cf. I. Warhaftig, "Haganat Hatzarkhan Leor Hahalakha," *Tehumin* 1 (1980), pp. 444, 468ff. If adjudication is his sole occupation, however, the judge is permitted to accept remuneration, as Asheri notes in commenting on bBekhorot 29a: "It is different if they devoted all their time to hearing cases, and could not engage in any other work, and they could not be expected to starve, and had to accept payment for their livelihood." Cf. *Tosafot* on bKetubot 29a s.v. *ma ani.*

Furthermore, Rabbenu Tam distinguishes between remuneration offered by the litigants, which judges are forbidden to accept, and that paid by the public purse, which is permitted (*Tosafot* on Ketubot 105a s.v. *gozrei gzeirot*). These distinctions validate the distinction I drew above (near note 42ff.) between two types of courts.

53 See e.g., bKetubot 106a s.v. *hai ase vehai ase*; Rashi *ad loc.* s.v. *hai ase*: "the duty of the judges is a positive precept: And thou shalt judge righteously [Deut. 1:16]."

Two vital and interrelated elements of any contractual obligation – that the obligation be new, rather than outstanding or previously incurred, and that the parties to the contract be at liberty to incur or decline the obligation – are thus absent in the case of courts.

5. An Argument Against the Analogy with the Laws of Agency

In addition to the conceptual difficulties with the analogy already noted, a further problem arises from the attempt to apply the laws of agency to the judge–litigant relationship.

According to Jewish law, an individual cannot judge himself. This is well-known from many pericopes and requires no further elaboration. Since the Sages disqualified themselves from judging anyone who had benefited them in any way,[54] they naturally disqualified themselves from adjudicating disputes in which they were involved. Asheri states: "Given that it is clear that one may not judge the case of a relative, obviously one is not to judge oneself."[55]

If one is not permitted to judge his own case, how can it be argued that a judge acts as agent of the litigants? As R. Huna b. R. Joshua states, "Is there anything that we cannot do and they can?";[56] likewise, the Gemara asserts: "It is said, one cannot appoint an agent except in matters with regard to which

---

54 bShabbat 119a, bKetubot 105b, bSanhedrin 7b.

55 Asheri on Moed Katan 3:10. Self-help is irrelevant in this context, because it does not constitute a conclusive judicial resolution of a dispute, but rather, an interim remedy for one of the prospective litigants pending and subject to court action and possible reversal. Final approval must be conferred by a court ruling to the effect that the remedy is lawful. Cf. S. Ettinger, "*Haose Din Leatzmo Bemishpat Haivri*," (PhD diss., Hebrew University of Jerusalem, 1982). For a thorough discussion of this subject, see E. Quint and N. Hecht, *Jewish Jurisprudence* (New York: 1984), vol. 2, ch. 4, and S. Ettinger, *Self-Help in Jewish Law*, vol. 6 of *Selected Topics in Jewish Law*, ed. H. Ben-Menahem and N. Hecht (Tel Aviv: 1993).

56 bKidushin 23b; compare bBaba Kama 110a on the case of an elderly, ailing priest: "If he cannot perform the priestly service, how can he appoint an agent?"

That one cannot appoint an agent to perform a task which one is himself legally incompetent to perform is such an immutable principle in the laws of agency that a Mishnah was rephrased so that it would be compatible with this principle. The Mishnah in question relates that on the Day of Atonement, the Elders say to the High Priest: "Eminent High Priest, we are the delegates of the court, and you are our representative" (Yoma 18b). The Talmud objects, on the basis of R. Huna's argument, that neither the delegation nor the court has the authority to perform

one is legally competent; in matters with regard to which one is presently legally incompetent, one cannot appoint an agent."[57]

These maxims imply that the applicability of the institution of agency is limited. One who experiences a practical difficulty which prevents him from conducting his affairs as he wishes may indeed invoke the institution of agency, but one lacking legal competence to perform an act cannot overcome this impediment through the institution of agency. In our context, this idea leads to the following conclusion: the legal restriction that one cannot judge one's own case cannot be circumvented by the appointment of an agent.

The attempt to regard the judge as an agent raises a further difficulty by contradicting the talmudic dictum, "An agent's acts are attributed to the principal." If the agent is indeed tantamount to the principal, then he is disqualified from judging the principal.[58] This problem also obtains with regard to the issue of "an agent who acts as a witness,"[59] and underlies the argument of those who claim that an agent cannot testify: he cannot testify for the principal because he is virtually equivalent to the principal himself. On this view, we cannot maintain that the judge is an agent of the litigant. While the Jerusalem Talmud states, "There is a Mishnah which does not accept the rule that an agent's acts are attributed to the principal,"[60] this only means that the principle attributing the agent's acts to the principal is not absolute, in that it does not disqualify the agent from testifying (and the judge from judging). Indeed, this is the rationale of those who say that an agent may become a witness.

The difficulty arising from the words of R. Huna b. R. Joshua[61] can be partially resolved by following a suggestion in bKidushin.[62] The Talmud states

Temple services; hence, they cannot delegate this power to the High Priest. The Talmud offers to change the text of the Mishnah from "you are our representative" to "we swear you in on our authority and on the authority of the court." Cf. R. Hananel's commentary *ad loc.* Nevertheless, the passage in bKidushin 41b appears to be unfamiliar with this principle.

57 bNazir 12a. I understand these two maxims as expressing the same idea, but cf. *Noda Biyehuda*, 2nd ed., Even Haezer, #70; *Mitzpe Eitan*, bKidushin 23b; *Bnei Ahuva*, Laws Concerning Marriage, 9:6.

58 Cf. bKetubot 105b: "The agent takes on a similar mentality, and by identification, becomes like the principal, and a man does not see his own faults"; and cf. bShabbat 119a.

59 Above near note 31ff.

60 jKidushin 2:1.

61 At note 56 above.

62 53a.

there that a Canaanite slave may appoint an agent to receive his bill of manumission, although he cannot receive this document directly from his master. This law contradicts the statement of R. Huna b. R. Joshua. The Talmud resolves the inconsistency by arguing that since the slave may accept a bill of manumission for another slave, one who belongs to another master, he is "generally competent with regard to such documents," and thus can appoint another to receive his bill of manumission for him, though he cannot receive it himself. This constitutes a limitation on the rule of R. Huna b. R. Joshua: it is not required that the principal be able to perform the action with which he charges the agent, but the principal must be able to perform other actions which fall under the same category as that with which the agent is charged.

Similarly, we can say that since any Jew is eligible to judge another, he is "generally competent with regard to judging," and therefore can appoint another to judge his case, though he himself is unable to do so. This is, however, only a partial solution.

In our context, the concept of being "generally competent with regard to (a given category)" can be interpreted in at least two senses. It could be said that we require of the principal that he be competent to perform the act in question at some point, even in the past or future and not necessarily at the time of appointing the agent. On the other hand, it could be argued that it is not enough if the principal was or will be competent to perform such an act at some point in time, but what is required is that the principal be competent at the time of appointing the agent.

Indeed, there is a discussion in the Talmud which indicates that the latter interpretation is correct – that the principal must be competent at the time of appointing the agent. The passage states that since an ailing priest is, at that point in time, legally unfit to discharge his duties, he cannot delegate them to others, in spite of the fact that he was fit to discharge them previously, and will be fit again upon recuperation.[63]

In light of this, it follows that the agency model will be unable to account for the judge–litigant relationship in cases where the litigant is a minor or is insane. Indeed, when such a minor grows up, or such a demented individual recovers, he is deemed "generally competent" to judge, but for the present, he is not competent to judge.[64] Furthermore, women are not able to become

63 bBaba Kama 110a.

64 See *Tosafot* on Gitin 22b s.v. *veha*.

judges: they are **always** denied the legal competence to judge.[65] Thus even if we adopt the first interpretation, the agency model will still fail to capture the relationship between a judge and a female litigant. The case of women therefore remains an anomaly on the agency model of the judge–litigant relationship.

6. The Judge as Agent of the Public

These difficulties impel us to consider the possibility of regarding the judge as an agent of the public. This strategy allows retention of the talmudic analogy as reflected in the various passages examined, while obviating the objections raised in the previous sections. In particular, the case of female litigants would no longer be anomalous. Furthermore, the model would be comprehensive enough to apply to courts of the second type delineated above, namely, non-voluntary courts. It is interesting to note that the first mention of the principle that an agent's acts are attributed to the principal in the Tannaitic sources refers to the classic example of an agent of the public – the synagogue reader.[66]

The agent of the public is regarded as a legal institution subject to the usual laws of agency. On this conception of the status of the judge, the judge has no particular dependence on any one side, as seems to be called for if the judge is construed as an agent of the litigant, but rather, his allegiance is to both parties to the dispute. This is emphasized in one of the laws referred to above: "[Sale by] the court is only void if it deviates from the market value by a sixth or more."[67] The Early Authorities raised the following question about this Mishnah: why is the sale void if the price was increased? After all, the judges benefited the person whose property they appraised. One interesting reply is as follows: "Just as the orphans rely on them, so does the purchaser, who says 'The court will do no wrong, and they do not deviate from the law out of pity,' for **they become agents of both sides**" (my emphasis).[68]

---

65 On the incompetence of women to judge see: Mishnah Nida 6:4, bBaba Kama 88a, jYoma 6:1, bGitin 5b: "sometimes a woman brings it [the bill of divorce]"; *Tosafot* on Baba Kama 15a s.v. *asher tasim lifneihem*; *Tosafot* on Jebamot 45b s.v. *mi lo tavla lenidota*; *Tosafot* on Nida 3a s.v. *kol hakasher ladun kasher lehaid*; *Tosafot* Gitin 88b s.v. *velo lifnei hediotut*.

66 Mishnah Berakhot 5:5.

67 See note 28 above.

68 *Shita Mekubetzet*, Ketubot 99b; also see Asheri and Meiri on Ketubot 100a, and Rabbenu Nissim on Alfasi *ad loc.*

The idea that the court acts in the interests of both sides is expressed by the Talmud as follows: "Law and charity – law for one and charity for the other; law for the one whose property was returned, charity for the one from whom the stolen item is taken to restore to its owner."[69] On this description, the act of adjudication is not understood as that of an impartial body charged with weighing the conflicting interests of adversaries who appear before it, but as an act undertaken to further a common goal – restoration of order and the distancing of the transgressors from iniquity (in this case, from theft). Thus both parties benefit from the judge's act.

The conception of the judge as an agent of the public obviates the difficulty arising from the view of the judge as agent of the litigant, namely, the fact that an agent is subject to the will of his principal and his authority to act is limited by the principal's instructions, while a judge is subject only to the rule of law.[70] Granted, this last statement must be qualified to accommodate voluntary judicial frameworks, wherein the litigant may stipulate conditions which must be fulfilled for the judge to be authorized to act.[71] From this point of view, voluntary adjudication conforms to the agency model. But with regard to compulsory adjudication, obviously there is no room for the litigant to specify the parameters of the adjudication or to set any conditions by which the judge must abide. This adds further support to the distinction between voluntary and compulsory adjudication proposed above.[72]

In any event, if we regard the judge as an agent of the public, then his subordination to the law raises no particular difficulties, since the law can be identified with the will of the public. On this conception, in acting as its agent, the judge is subordinated to the wishes of the public, his principal.

Various statements in the Talmud support the idea that judges are agents of the public. A good example is the following assertion: "When you see a generation visited by much calamity, go and examine the judges of Israel,"[73] which specifies a definite correlation between the public and its judges. The incidence of misfortunes indicates that the judges lack integrity, for which the

69 bSanhedrin 6b.

70 Cf. G. Tedeschi, "Apotropos Veshaliah," in *Masot Bemishpat* (Jerusalem: 1978), especially p. 339.

71 See, e.g., bSanhedrin 5a: "if they [the litigants] submit to your authority, you are not obliged to pay [for your errors]." Here, the litigants set the parameters of the judge's responsibility.

72 See above near note 42.

73 bShabbat 139a.

public is punished. A similar idea is found in the Mishnah: "One who makes an error in his prayers, it is a bad sign for him; if he is the public reader, it is a bad sign for those who appointed him, because an agent's acts are attributed to the principal."[74] Above, we saw examples where the agent's acts benefited his principals; here, the agent – the reader – suffers for the sins of his principal. This association between the public and its judges is also hinted at in the following statement: "The public is cautioned to respect judges, and the judges are cautioned to be tolerant with the public."[75]

The attempt to view the judge as an agent of the public is apt to raise the question of whether a judge can adjudicate in a dispute between the public and an individual when the question of partiality arises. The problem is discussed in the following *baraita*:

> Residents of a town whose Torah scroll was stolen – the case is not tried by the judges of that town, and testimony by townspeople is not admissible.
> One who said: "Give a *mane* to the people of my town" – the case is not tried by the judges of that town, and testimony by townspeople is not admissible.[76]

This *baraita* is quoted by the Gemara in the context of a discussion about whether one can renounce one's financial ties to the subject of one's testimony in order to qualify as a witness. The Talmud reports on the authority of Samuel that business partners may testify for each other if the partner who is to testify forswears the partnership before giving evidence.[77] The *baraita* apparently does not consider such dissociation feasible, since it does not propose it as a solution to the problem of partiality that arises in such cases, and instead states the law categorically.

The *baraita* disqualifies townspeople from serving as either witnesses or judges in a dispute between the public and an individual. While this disqualification from testifying may be amenable to remedy by dissociation in those cases where such dissociation is technically possible, this solution does not suffice to overcome the disqualification from judging. For if judges are considered agents of the public, then even after forswearing any financial

74 Mishnah Berakhot 5:5.

75 bSanhedrin 8a; the version cited above is particularly interesting in light of the following variant: "The litigants are cautioned to have respect for the judge, and the judge to be tolerant of the litigants" (*Sheiltot* on Exod., 58).

76 bBaba Batra 43a; Tosefta *ad loc.* 9:9.

77 *Ibid.* 42b–43a.

involvement with those connected to the case being tried, as members of the community, they still remain interested parties, and as such are disqualified. We faced the same problem in considering the judge an agent of the litigant, which brought to mind reservations raised in the passage on the "agent who acts as a witness."[78] To be sure, the Talmud concludes that an agent may act as a witness, so it could be argued that likewise, the judge's serving as agent of the public does not disqualify him from adjudicating. But the question of whether citizens of a town can serve as judges after dissociating themselves in the proposed manner is not conclusively resolved, in spite of the fact that the *baraita* implies that the townspeople's incompetence to serve as judges is perfectly analogous to their incompetence to serve as witnesses, which, we noted, can be overcome.

The Talmud's approach to this *baraita* raises a difficulty much discussed by the Early Authorities. In articulating the problem to which the *baraita* gives rise with regard to the feasibility of qualifying for testimony by dissociating oneself from one's financial interests, the Talmud states: "Let two of them declare that they have no share in the matter, and then they can adjudicate."[79] But this is perplexing. For a court is constituted by three, not two, though two suffice as witnesses. Therefore, we would have expected the Talmud to assert: "Let two of them declare that they have no share in the matter and give evidence," or "Let three of them declare that they have no share in the matter and then they can adjudicate."

The following solutions, among others, have been proposed: (1) the Talmud's statement is elliptical, though its full meaning is indeed as suggested above;[80] (2) the text should be emended to read: "Let two of them declare that they have no share in the matter and then they can testify";[81] and (3) the Talmud here reflects the approach of Samuel, who contends that "two who judged, their ruling is valid, but they are considered an insolent court."[82]

Nachmanides' explanation merits special consideration. "Let two of them declare that they have no share in the matter and give testimony, and they will be judged before three others"[83] – others who are not chosen from among the townspeople. That is, the passage in the Talmud relates only to the

78 See above near note 54ff.
79 *Ibid.* 43a.
80 See the commentaries of the Rashba, Ritba and Ran *ad loc.*
81 *Tosafot Rid ad loc.*
82 *Yad Rama ad loc.*; cf. Shakh on *Shulhan Arukh*, Hoshen Mishpat 37:32.
83 Ramban *ad loc.*

dissociation of witnesses, in line with Samuel's statement with regard to business partners who forswear their association and then testify. After the witnesses have renounced their interests and testified, a court in another locale can adjudicate the case on the basis of their testimony. This interpretation, which is supported by the variant reading in the text of the Hamburg Manuscript [of the Talmud] – "and we shall judge them" – thus precludes the possibility of judges abjuring their financial interests in order to judge their fellow townspeople, and hence is compatible with the notion that judges are agents of the public.

7. The Judge as Agent of the Almighty

A passage in the Talmud on the status of priests, which promises to shed new light on the matter at hand, has yet to be explored. The passage in question is the following: "The priests – are they our agents, or agents of the Almighty?"[84]

The very fact that this question is articulated at all indicates a connection between the problems I have identified with respect to the status of judges and an explicit concern of the Talmud. I refer to the discussion about whether the status of a priest who performs a service for the public is that of an agent, since if this is the case, the laws of agency apply to priests. The expression, "agents of the Almighty," is undoubtedly a metaphor arrived at by way of contrast with the alternative, "our agents," because obviously it is not possible to describe the link between the Almighty and the priests as a legal relationship in the strict sense. Rather, a metaphor is used by the Talmud to express the idea that the status of priests originates from a divine commission, and hence the relationship between the priests and the rest of the nation is not an agent–principal relationship.[85]

The Gemara considers the question at bNedarim,[86] but does not resolve it; in bKidushin, however, a conclusion is indeed formulated. Given the argument, "Is there anything that we cannot do and they can?,"[87] the attempt to apply the laws of agency to priests makes it clear that priests cannot be

---

84 bNedarim 35b, bYoma 19a, bKidushin 24b.

85 Cf. M. Silberg, *Talmudic Law and the Modern State*, trans. B.Z. Bokser (New York: 1973), pp. 1ff., especially p. 9; id., *Beein Keehad* (Jerusalem: 1982), pp. 160ff., especially p. 170.

86 See note 84.

87 bKidushin 23b.

agents. From the fact that the laws of agency are inapplicable to priests, the Talmud concludes that priests are agents of the Almighty.

This pericope lends credence to the hypothesis that it may be possible to define the status of judges in a similar manner, since the problem which prevents us from regarding the priests as agents of the rest of the nation also arises with respect to judges. I do not claim that the discussion in Nedarim encompasses the problematics ensuing from the attempt to clarify the status of judges, though this is not impossible. From a historical point of view, there is a definite connection between the two institutions.[88] Here, however, I argue only that it may be possible to extrapolate from the passage in Nedarim in order to accurately gauge the status of judges.

That judges are regarded as the agents of the Almighty, in the metaphoric sense, can be ascertained from a number of Tannaitic dicta expressing the view that "the judgment is God's" [Deut. 1:17] and, to use the words of R. Abraham Ibn Ezra, that judges "uphold divine law on earth."[89] Judges have thus been charged with a special responsibility, so to speak, by God, and if they neglect their duty, they impose upon Him the task of rectifying what they have thwarted.[90]

Against this background I see more than just a manner of speaking in R. Moses Sofer's assertion, regarding the basis of judicial authority, that "All those who are fit have been ordained by Heaven in any case."[91]

The idea that the judge is a divine agent is put forward explicitly in Nachmanides' comments on the verse, "Ye shall not be afraid of the face of any man; for the judgment is God's":

> It is for God to administer justice among His creatures, for He created them to have equity and justice among them, and to save the victim from his abuser, and He placed you [the judges] in His stead. If you will be partial and perpetrate

---

88 Cf. Deut. 17:9; Sifre Deut. 153, pp. 206–8; Tosefta Sanhedrin 4:7; bSanhedrin 52b. Also see M.D. Her, "Haretzef Shebeshalshelet Mesirata Shel Hatorah," *Zion* 44 (1979), pp. 43–56.

89 Ibn Ezra, Commentary on Exod. 21:6. Biblical verses usually cited in this context include Prov. 22:22 and II Chron. 19:6 as well.

90 See, e.g., bSanhedrin 8a: "it is not sufficient for the wicked to take money from one and give it to others [but they force Me to intervene]" (*Sheiltot* on Exod., 29, reads "judges" for "wicked"); jSanhedrin 1:1: "The Holy One said: I told you to give it to Reuben, but you take it from one and give it to another."

91 Hatam Sofer, *Responsa*, Hoshen Mishpat #163.

> injustice, behold, you have sinned before Him because you have betrayed your mission.[92]

8. Conclusion

Having come full circle, it is now time to assess the judge–agent analogy. In spite of its creative contribution to the derivation and formulation of a number of important laws, there is no true analogy between the judge and the agent. The apparent analogy is in essence little more than an evocative turn of phrase. The view championed by the Hatam Sofer, then, seems to be the correct one.[93]

Two factors facilitated the emergence of the purported analogy: the Talmud's wide use of the institution of agency to clarify legal relationships, on the one hand, and the rather ambiguous notion of judges as agents of the Almighty, on the other.[94] Both factors are reflected in the following quotation from Sifra:

> That which the Lord commanded Moses – the agent is worthy of his Principal; that which Moses commanded the children of Israel – the agent is worthy of the addressee, and the addressee worthy of the agent.[95, 96]

92 Nachmanides on Deut. 1:17.
93 See above at notes 4 and 5.
94 See the preceding section.
95 Sifra, Behukotai, at the end of the passage.
96 This passage also vividly captures the principal–agent–addressee relationship.

CHAPTER 3

THE LEGAL STATUS OF THE RESPONSA LITERATURE

Berachyahu Lifshitz

1. Introduction

> The responsa literature is the third major type of literary source of Jewish law in the post-Talmudic period.... The responsa literature is unique in many respects: its essential character, which is unlike the other types of post-Talmudic halakhic literature; its subject matter; its importance for *mishpat ivri* in particular, as well as for Jewish law in general; its significant contribution to the history of the literary sources of the various periods of Jewish law; and the rich storehouse of information it provides about the history of the Jewish people in its different centers.[1]

These are the opening words of Elon's comprehensive study of the responsa literature. Elon discusses each of these features in turn. The responsa literature

---

1 M. Elon, *Jewish Law* (Philadelphia and Jerusalem: 1994), vol. 3, pp. 1454–1456. See Elon's introduction to the *Digest of the Responsa Literature of Spain and*

is a prime source of information for the various sub-disciplines within Jewish studies;[2] and one cannot exaggerate the importance of its contribution to the development of Jewish law.

However, the status of this literature in the Jewish legal system is not unequivocal. Elon states that

> Clearly, the authority of the respondent stemmed mainly from his expertise in the *Halakhah* and from his spiritual stature. The writings of some authorities were viewed in many places, even in their own lifetime, as binding... such personages were viewed as authoritative respondents. However, as pointed out... the halakhic basis for the authority of a respondent still has not been completely clarified.[3]

Yet this is not so certain. Indeed, it seems that the authority of these scholars was not a direct outcome of their greatness and personality, though these attributes obviously contributed to the litigants' and petitioners' acceptance of their rulings as binding and to the acknowledgement of their authority by other halakhic scholars, both in general, and with regard to the cases at issue. Rather, their authority is based on their being accepted as judges by individuals and the community. The fact that an individual who acknowledged the authority of a given respondent was impressed by his greatness, his fame, his literary style or any other factor, is, in principle, of no importance. Naturally, the opinions of great scholars of outstanding spiritual stature were regarded as more binding than those of respondents of lesser renown. We also have to distinguish between claims to authority made by certain scholars, and actual acknowledgement by the community. The two are not necessarily commensurate: some scholars, out of genuine humility, thought themselves unworthy of

---

North Africa (Hebrew), (Jerusalem: 1981), Mafteiah Hamekorot, vol. 1, pp. 9ff. Also note the following:

> Geonic responsa were termed "questions" (more precisely, "questions and answers") [sheeilot uteshuvot] here, and also in other Geonic works of the early and late periods and in the post-Geonic literature... . "Questions," rather than "questions and answers," or "answers," was thus the accepted term for responsa already in the middle Geonic period [R.J.N. Epstein, *Peirush Hageonim Leseder Toharot* (Jerusalem-Tel Aviv: 1982), pp. 69–70].

2 In fact, not only in Jewish studies, since the responsa literature contains information about economic conditions in various countries and periods, legal institutions, language, customs and so forth; see Elon, *Jewish Law*, vol. 3, p. 1463.

3 Elon, *Jewish Law*, vol. 3, p. 1467, note 43.

ruling on legal matters, but were nevertheless recognized as the leading authorities of their times, while others clamored for recognition to no avail.[4]

R. Benzion Uziel, the late Sephardic Chief Rabbi, writes:

> The books of responsa show that the arguments of both parties were forwarded by a court or by an officiating judge to the rabbinical authorities in Jerusalem who bore the title "Foremost in Zion" *(rishon letzion)* and whose scholarship befitted their lofty office. They constituted the Great Court recognized by the entire Jewish people. Their adjudication was sought through correspondence – the arguments of both sides were presented before them, and they were asked to decide. There was no appealing their rulings, nor were they open to criticism. This also describes the status of other great scholars of the Diaspora, who, by virtue of their wide knowledge and great wisdom, were acclaimed by their contemporaries and recognized as the paramount authorities of the generation. The public turned to them with questions, and litigants from all over the Diaspora who wished their cases to be tried by a Great Court or by a court of experts sent in their arguments.[5]

This account is not entirely accurate, and is best understood as an idealization. The attempt to construe respondents as a Great Court or a court of experts[6] is apparently based neither on historical fact nor on substantive legal reasoning. All we can say about those respondents who were recognized as legal authorities by a great part of the Diaspora community, is that even in the absence of formal authority (such as that of a Great Court or other courts with the power to enforce their rulings), to some extent their actual conduct resembled that of a Great Court.

Elon, in commenting on this matter, is cautious:

> In the Jewish legal system, there was always a supreme tribunal which dealt with problems the lower courts were unable to resolve, or to which certain types of legal issues or cases involving special circumstances were submitted in the first

---

4 According to R. Joseph Caro, submission and acceptance are the source of the authority of the Babylonian Talmud as well; see *Kesef Mishne*, Laws Concerning Rebels 2:1. They are also the basis for the authority of the community and its representatives. Even if there is an element of fiction in the assumption that such acceptance exists, the very fact that it is made indicates its importance. Cf. Elon, *Jewish Law*, vol. 3, pp. 1459–1460, note 20; and see A. Schiller, "Jurists' Law," 58 *Columbia L.R.* (1958), p. 1236, regarding the source of the Roman jurists' authority. (I thank Prof. A.M. Rabello for bringing this to my attention.)

5 R. Benzion Uziel, *Mishpetei Uziel* (Tel Aviv: 1940), part 3, Hoshen Mishpat, sec. 1.

6 For explication of these terms, see *Encyclopedia Talmudit* s.v. *bet din.*

instance. In the Geonic period, such supreme tribunals existed in the Land of Israel and in Babylonia. The academy *(yeshivah)* in the Land of Israel considered itself to be "the court for the entire Jewish people." The patriarch in the Land of Israel declared that his jurisdiction extended to the entire Jewish people, even to those residing in far-off places, and that judges everywhere were subject to his authority. The Babylonian academies – in theory and in practice – regarded themselves as the successors to the High Court in Jerusalem, with all of the prerogatives that flowed from that status.

With the decline in the authority and influence of these centers, the respondents, who were the legal and spiritual leaders of their generation – increasingly functioned as supreme tribunals. Complex problems that local courts could not resolve, and fundamental questions transcending the immediate parties and involving the wider public, ultimately reached one of the recognized respondents active in the various centers in the diaspora. The respondents constituted an unofficial and informal supreme tribunal for Jewish law. In this role, the respondents also contributed to the preservation of a certain measure of uniformity of decision and to the maintenance of close ties among the different communities with each Jewish center and between one center and another.[7]

This perception of the post-Geonic respondents as a quasi-supreme court is informed by hindsight. The respondents themselves made no claim to this status, but rather, acknowledged their lack of formal authority. In this they differed from the Geonim, for not only were the Geonim highly conscious of their status, they also did not hesitate to assert their authority, even when it was challenged. But as time passed, any resemblance to a formal judicial system gradually diminished, and institutions with actual power were replaced by more abstract frameworks. Thus there is reason to dispute the thesis of the essential continuity of the later responsa with those of the mishnaic, talmudic and Geonic periods. The distinguishing feature, as we noted, lies in the different status of the respondents in the two periods. The distinction is illustrated by examining the respondents' practice of answering questions addressed to them by one side only in disputes relating to monetary matters.[8] This analysis is most illuminating, for it suggests a criterion for defining the essence of the responsa literature.

---

7 Elon, *Jewish law*, vol. 3, pp. 1460–1461.

8 It is also discussed in Elon, *Jewish Law*, vol. 3, pp. 1501–1506, and E. Shochetman, "The Obligation to State Reasons for Legal Decisions in Jewish Law" (Hebrew), *Shenaton Hamishpat Haivri* 6–7 (1979–1980), pp. 338ff. But further study is required.

2. Responsa Addressed to Only One of the Sides

Many respondents were aware of the prohibition against rendering an opinion if requested to do so by only one of the parties to the dispute. This prohibition was manifested in a variety of forms, from expressions of displeasure toward those who rendered such opinions, to issuing enactments, backed by sanctions, bluntly prohibiting such conduct. A number of different rationales for the prohibition were offered, some explicitly, others tacitly; similarly, various rationales were given for either enforcing the prohibition or ignoring violations. As will be shown below, the various rationales were not necessarily compatible with each other. Never articulated as a consistent set of arguments, as the popularity of one claim declined, another emerged in its place. Let us review how these arguments unfolded.

i. "Take Not the Counsel's Part"

The first argument against responding to one of the parties to a dispute without the agreement and representation of the other is that by conducting himself in this way, the respondent in effect makes himself a counsel for one of the parties, thereby transgressing the following mishnaic injunction: "Judah b. Tabai says: Take not the counsel's part."[9] Various suggestions as to the meaning of "counsel" have been offered. Most commentators agree that a counsel is one who advises and represents litigants; they disagree only over details.[10]

In a responsum by R. Yehudai Gaon, we find the following statement:

> He was asked the meaning of "take not the counsel's part." He told them: A case which they do not want to decide before you and one of the litigants comes and says: I want to set out this case before you. Do not listen to him.[11]

---

9 Mishnah Avot 1:8.

10 See J.A. Blau, "Hashvaa Bein Hamishpat Haivri Lahellenisti-Mitzri,"*Kovetz Hamishpat Haivri* 5 (1937), p. 14. Blau suggests a different reading, replacing the term for "counsel," *orkhei dayanim*, with *arkhei dayanim*, or "head judge[s]," which is the original sense of the term. Also see B. Lipkin, "Arikhat Din Bemishpat Haivri," *Sinai* 30 (1952), p. 46; 31 (1952), p. 265; S. Assaf, *Batei Hadin Vesidreihem Aharei Hatimat Hatalmud* (Jerusalem: 1924); N. Rakover, *Hashlihut Vehaharshaa Bemishpat Haivri* (Jerusalem: 1972); H. Tchernowitz-Rav Zair, "Hasanegoria Bebatei Dinin Shel Yisrael," *Hashiloah* 3 (1898), pp. 422–481.

11 B.M. Lewin, *Otzar Hageonim*, Ketubot (Jerusalem: 1939), Helek Hateshuvot, sec. 349.

According to this version of the Geonic responsum, we do not know why the judge should not listen to one who asserts that he is not seeking an actual verdict; that is, it is unclear why such an individual should not be able to state his case before the judge. However, a fuller version of this responsum is quoted in Rashi's commentary to the passage from Mishnah Avot.

> And R. Yehudai Gaon of blessed memory said about another matter: If someone appears before you and says to you, "I do not ask you to judge for me, or to be a court for me. I am only asking you the law regarding the demands and claims I am making against this man." You must not answer him about the law or explain to him anything related to it, because this is the way of Gentile judges: people come to ask them the law and they reveal it.

The argument here is that the scholar must not give his opinion about the law as applied to a given case, because offering advice of this kind is the custom of the Gentiles. The petitioner's disclaimer, "I do not ask you to judge for me, or to be a court for me," is emphasized. It is implied, therefore, that had the petitioner stated that he was willing to accept the scholar as a judge and the decision as binding, there would have been no infringement of the injunction "take not the counsel's part."[12] In other words, one whose authority to rule on the case is accepted by the petitioner is permitted to give his opinion on the case, but one who is not going to decide the case is not permitted to do so.[13]

R. Yehudai Gaon's comments help explain a passage in the Jerusalem Talmud which commentators have found problematic. The passage is as follows:

> Like the kinswoman of R. Simon b. Va who suffered from an eye ailment. She came before R. Johanan and he told her: Is the wage of the surgeon fixed? If it is fixed – from your marriage contract; if not fixed – your husband pays it. But did we not learn differently? "Take not the counsel's part"; R. Haggai said in the name of R. Joshua b. Levi, "Should one not conceal the outcome of the case from

12 Whether a judge is allowed to hear one of the sides in the other's absence is a different question. It is discussed below, but it can already be stated that the prohibition of such behavior on the part of the judge seems to be related to the fact that the judge is responsible for establishing the facts of the case. Were he to hear one side in the other's absence, he might be inclined to favor that side's version of the events in question.

13 There seems to be no need to seek the other side's agreement to be judged by the scholar in question. Apparently, it suffices for the petitioner to express his own willingness to be bound by the ruling, though it is not binding on the other side. See the text at note 15.

> the individual?"; We answer: R. Johanan knew that she was a respectable woman and that is why he told her. If her husband prefers the surgeon with the fixed wage and she prefers the one with the variable wage, whom do we heed? Do we not accept the husband's preference? R. Matanya said: That is to say, when the law is not with him [one may not reveal the law], but when the law is with him, one may tell him the law.[14]

What is relevant to our concerns is the conclusion. The *Pnei Moshe* commentary interprets it as follows:

> It is forbidden to answer only when the judge does not know that the law is with him but says: "if the case were so and so, then the ruling would be like this," and through this the petitioner learns a useful line of argument. But if the judge knows that the law is with him and he only asks if the law is according to his claim, then the judge may tell him. He tells him "you are right" because there is no suspicion here that he will learn a new line of argument from the advice, because the litigant is merely told that according to his argument he is right.

This interpretation, however, is unclear. What exactly is the difference between the two instances? The *Zion and Jerusalem* commentary connects the passage from the Jerusalem Talmud with the comments of R. Yehudai Gaon: "We can interpret the Jerusalem Talmud as saying that the restriction applies only when the other party is absent, but when the other party is present they may be told."

Though I agree that the Jerusalem Talmud should be interpreted using the comments of R. Yehudai Gaon, it should not be understood in the manner proposed by *Zion and Jerusalem.*[15] On my interpretation, if the petitioner will not be appearing before him for judgment, a judge is forbidden to tell him anything. But if the petitioner will be appearing before him, it is permissible to reveal the outcome to him.

The passage in the Jerusalem Talmud has a parallel version in the Babylonian Talmud, in which the narrative is essentially the same.

> Kinsmen of R. Johanan had a stepmother who needed medical care. They came before R. Johanan every day. He told them: Go and set a fixed wage for the surgeon. R. Johanan said, We took the counsel's part. What was his reasoning

---

14 jBaba Batra 9:6, 17a; the parallel version in jKetubot 4:11, 33a, is abridged at the end.

15 See note 13 above. For the text does not say "when the litigant is with him" (*baal dino imo*), but rather, "when the law is with him" (*dino imo*).

> before and what was his reasoning after? Before he thought, "Do not ignore thy kinsman" (Isaiah 58:7), at the end he thought, I have been important here.[16]

The Geonim R. Sherira and his son R. Hai were asked to explain this passage:

> And what you ask: Why did R. Johanan say this? Was he not then the head of the Academy and were not all cases decided according to his opinion?[17]

They replied:

> The law is according to R. Simon b. Gamaliel concerning a widow in need of medical care. If the treatment has a fixed charge she claims it from her marriage contract. But these people did not come for a ruling and R. Johanan did not adjudicate for them. And if he had been silent and had not told them anything, they would have paid the medical expenses from the estate of the deceased as part of the widow's maintenance. But through R. Johanan's advice, they caused her a loss, because the ruling that the expenses were deductible from the marriage contract is a new ruling. Since he taught them a remedy against her arguments he was giving advice for the gain of one at the expense of another, which is acting like a counsel, and the rule is "Take not the counsel's part." For R. Johanan did not serve as a judge for them.[18]

In this passage, too, it is stressed that one who gives an opinion should do so as a judge, otherwise his answer constitutes "taking the counsel's part." But the respondents have another reservation: the response may cause a loss for the opposing party. It seems that the reasons given by R. Sherira and R. Hai differ from that cited by R. Yehudai Gaon, who does not distinguish between an opinion which causes the other side a loss and one which does not. On his view, it will be recalled, the basic reason for the prohibition is that

---

16 bKetubot 52b.

17 *Otzar Hageonim*, Ketubot *ad loc.*, sec. 348.

18 The respondents add, in explanation, that at the outset R. Johanan did not want to do something he would later regret, but mistakenly thought that one should not ignore his relatives; later, it was this that he regretted. This case is similar to one reported in bKetubot 86a. But cf. responsum #345 in R. Joseph Haim's *Tora Lishma* (Jerusalem: 1973), where he rules, on the basis of bSanhedrin 27b, that one who writes an opinion in order to help one of the litigants is not allowed to take payment for his efforts because he is considered as one of the judges deciding the case.

the conduct in question resembles the ways of the Gentiles. R. Sherira and R. Hai, however, do not consider the ban on "resembling the Gentiles"[19] to be the reason for the prohibition; indeed, they do not even mention it as a relevant factor. In any case, both sources agree that "counsels" are those who do not serve as judges in the cases in question.

It appears that the responsum of R. Sherira and R. Hai can explain one of Rashi's responsa:

> Heavenly peace shall comfort you, Isaac the Orphan, and establish your suit on truth and peace. We the signatories reply about your case; we do not act the counsel's part and provide a definitive judgment to one of the sides. We do not want to transgress the interdiction of "uttering a false report."[20] However, to act so that one may keep what is rightfully his, which divine providence made his inheritance, until a lawsuit is brought with arguments for taking it from him, in this there is no taking the counsel's part. Therefore we give you a brief account of what to do.[21]

To justify their telling the petitioner "what to do," Rashi and the other respondents argue that in so doing they will ensure that "one may keep what is rightfully his ... [which is not] taking the counsel's part." This argument complements that of R. Sherira and R. Hai, who equate taking the counsel's part with causing a loss to be incurred by the adversary of the petitioner.[22] The same principle underlies both rationales.[23]

A similar argument can be found in a responsum by R. Solomon b. Simon Duran (Rashbash):

> And the way which I havé kept hidden is well-paved for those who understand, but I shall not reveal it to you, because you do not turn to me for a decision but

---

19 It is possible, though nowhere explicitly stated, that since the matter involves no less an authority than R. Johanan, who is described as "head of the Academy, in accord with whose opinion all cases are decided," it makes no sense to explain the prohibition in terms of similarity to Gentile practices.

20 The respondent appears to make a mistaken analogy here. The prohibition of assuming the counsel's role is not based on "Thou shalt not utter a false report" (Exod. 23:1), from which the Talmud (bSanhedrin 7a) deduces that a court must not hear one of the litigants before the other appears in court, and that a litigant must not make a statement before his opponent appears. See the text at note 60.

21 Rashi, ***Responsa*** (Elfenbein edition), #74.

22 As is pointed out by R. Yehudai Gaon above.

23 See L. Ginzberg, *Ginzei Schechter* (Newark: 1929), part 2, p. 47, and the sources cited there; and see the text below.

> for information, and I cannot see any way to reveal it without taking the counsel's part. For truly the counsel's part is to teach arguments to those who did not come up with them. And not as you understood, that the respondent who instructs the petitioner on how to act is taking, heaven forbid, the counsel's part. Because we find that the Sages of the Talmud did this. And the source of your mistake is a statement made in a responsum by R. Isaac b. Sheshet [Ribash],[24] and it is not as you thought... and even if it were as you thought, in our times I do not say it is merely permissible but that it is an important obligation, for it is like returning lost property. For I was alarmed to hear of the judges in those places, ignoramuses, self-styled arbitrators, and judges whose appointments came about through the governor's patronage, who judge for money and bend the facts. It is fitting that those who are able to render an opinion answer the petitioner so he does not perish without a trial. And I do not want to loosen the reins of my pen but rather to restrain my words so that they will yield peace.[25]

On the one hand the Rashbash defines "taking the counsel's part" as suggesting arguments the petitioner would not have thought of, tempting him to lie[26] in order to win his case. Yet he declares that even were he to agree with his inquirer that the prohibition should be broadened to cover cases where

---

24 The reference is apparently to R. Isaac b. Sheshet Perfet's responsum #179:

> For one who takes part in a dispute and responds to a matter without hearing the arguments of both sides together, is witless and dishonorable. ... further, from his answer, the litigant may learn a line of argument which might help him, and it transpires that the respondent has assumed the role of counsel.

25 R. Solomon b. Simon Duran, *Responsa*, #529.

26 The other arguments mentioned will be discussed below. In another responsum (#5), the Ribash restates the argument in responsum #179 as follows: "There is also the apprehension that the responsum will teach him to lie." The fear that litigants will learn to lie is palpable, and it is possible that the Ribash hinted at this concern in responsum #179, though he did not explicitly state that the claims the litigants would be advised of would be false (cf. responsum #175). Thus in his responsum R. Solomon b. Simon Duran reveals the legal point which he wanted to conceal from the general public and concludes:

> And this is a well-paved but hidden path, therefore I was right to say that I did not want to reveal this law, to avoid potential corruption. Were I not concerned that you would think me an ignoramus, I would not have revealed it at all. However, you should keep this secret and not reveal it to anyone, lest they learn to lie from it.

there is no fear of teaching the petitioner to lie, nevertheless, when there is a need to act so that a petitioner will not lose his possessions due to unlawful actions on the part of other judges, providing legal counsel is permissible. His attitude, then, is in line with that of the Geonim and Rashi.

The Rashbash thus restricts the scope of the prohibition, an attitude that is readily understandable if we keep in mind that the Rashbash and scholars who preceded him no longer thought of their responsa as judicial rulings, but rather, viewed them as learned opinions provided in response to inquiries. Given that the smaller the scope of the prohibition, the more room the respondents had to justify their involvement, inevitably, the charge that the respondents were acting as counsel was frequently levelled. There is thus a fundamental distinction between the responsa of the mishnaic, talmudic and Geonic periods, and those that followed, a distinction that plays a critical role in demarcating the role of the responsum in the Jewish legal system.

There are also minor differences that followed from the sense of authority which the earlier respondents enjoyed, and the later ones lacked, such as differences in the wording and scope[27] of the responsa, and in the ways in which the decisions are defended.[28] But these are merely secondary effects. The principal determinant of the nature of the responsum was undoubtedly the real authority of the respondents, and not merely their perceptions of that authority.

ii. Responsa Literature and Similar Institutions in Other Legal Systems

How did the loss of authority affect the development of Jewish law? Before addressing this question, it will be useful to reconsider the implications of R. Yehudai Gaon's opposition to responsa that do not constitute actual decisions. As noted above, he opposed the practice of issuing such responsa on the grounds that it resembled the approach of Gentile legal officials. R. Yehudai Gaon lived in Babylonia in the latter half of the eighth century. The area was, by that time, under Islamic domination. We therefore have reason to believe that he was referring to the practices of Moslem legal officials. It is known that, with respect to theoretical legal questions, the institution of responsa existed in the Islamic legal system.[29] Initially, the writing of responsa was an independent profession, open even to women, slaves, the blind and

---

27 Elon, *Jewish Law*, vol. 3, p. 1474.

28 Shochetman, *op. cit.* (note 8 above), p. 340.

29 Responsum – *fatwa*. See "*Fatwa*" in Schacht, *Encyclopedia of Islam* (1965), p. 866. Such responsa were brief and answered queries with a simple "yes" or "no."

the deaf, none of whom could serve as judges. With the passage of time, however, those who composed such responsa acquired an official status and were known as *mufti*. Though they performed a service for individuals, they were considered public officials. It became an accepted and familiar form of procedure, sanctioned by the judicial establishment, for the *mufti* to prepare opinions on legal matters which were then used by judges as the basis for their rulings. R. Yehudai Gaon opposed the attempt to reshape the practice of issuing responsa along the lines of the Islamic institution of the *fatwa*.[30] This unequivocal disapproval is particularly noteworthy in view of the fact that R. Yehudai Gaon wrote the earliest known responsa of the Geonic period.

In comparing the two types of legal opinions under discussion – the *fatwa* and the responsum – it is thus important to distinguish two different kinds of similarity. Historically, the two institutions developed during the same period, their purposes were somewhat similar, and there was some degree of mutual influence. However, these external or formal similarities are salient only from the standpoint of the historian. From the standpoint of the legal system itself there is no similarity, aside from some common terminology. In Jewish law the responsum is a judicial ruling, while in the Islamic system it is a learned opinion. Indeed, Jewish law opposes and strictly forbids the rendering of such opinions.

This same distinction should be used to evaluate the validity of the comparison between the responsa literature and the *responsa prudentium* of Roman law. Under the emperors Augustus and Hadrian, eminent legal scholars were granted official "authority to answer": they were licensed to write out answers to legal questions and imprint them with a special seal. These responsa were binding on the judges adjudicating the cases they referred to.[31]

---

30 Elon, *Jewish Law*, vol. 3, p. 1466; S. Baron, *A Social and Religious History of the Jews* (New York: 1958), vol. 6, pp. 109–110. Also H.Z. Hirshberg, "Hayehudim Beartzot Haislam," in his *Prakim Betoldot Haaravim Vehaislam* (Tel Aviv: 1968), p. 284; H. Lazarus-Yafe, "Leofya Shel Hatarbut Haaravit," in *Hartzaot Bekinsei Haiyun Behistoria* (Jerusalem: 1973), p. 181; S.D. Goitein, *A Mediterranean Society* (Berkeley: 1967), vol. 2, pp. 325–326, 598 (note 66); J.I. Bloomberg, *Arabic Legal Terms in Maimonides* (PhD diss., Yale University, 1980), p. 41. The *fatwa* undoubtedly influenced the responsa literature, and perhaps the reverse is true as well. It is also possible that the distinction which the Gaon was anxious to preserve disappeared over time; my intention is to identify the origins of this process.

31 W. Buckland, *A Textbook of Roman Law* (Cambridge: 1932), pp. 21–25; H.F. Jolowicz, *Historical Introduction to the Study of Roman Law* (Cambridge: 1972); R. Katzoff, "Responsa Prudentium," in *Roman Egypt, Studi in Onore di*

These Roman responsa are thus quite similar to the *fatwa*.[32] Here, however, we are only concerned with those points on which the two institutions differ from that of the halakhic responsum up to the close of the Geonic period.

At the same time, given the distinction made above between substantive and external similarity, it is also useful to evaluate the degree to which halakhic responsa are comparable to court rulings in other legal systems. According to Elon,

> The compilations of responsa in Jewish law are comparable to collections of judicial decisions in other legal systems, although there is a difference between them. Judicial decisions in other legal systems are rendered in cases that the parties have submitted to a court for determination, whereas a responsum is not handed down by a formal judicial tribunal. The respondent is an outstanding and recognized halakhic authority, and, generally speaking, it is not the parties but the local judge before whom the parties have appeared, who submits the question to the respondent.
>
> However, this difference between responsa and judicial decisions is only superficial and formal. The common feature of both is that both contain a description of the facts, the parties' arguments, the evidence, and the legal analysis; both include a legal conclusion and a decision, and both are authoritative and definitive – the one by virtue of the judge's formal legal authority, and the other by virtue of the respondent's spiritual eminence as a halakhic authority.[33]

However, this is an oversimplification. We have to distinguish between different periods. Until the end of the Geonic period there is both formal and substantive similarity between the two systems, since in both proceedings those who preside are official judicial authorities. After the Geonic period, however, the picture Elon portrays is only partially accurate. The factual descriptions and the formats of the different proceedings are indeed similar, but these are purely external features; Elon does not adequately stress the one truly salient formal difference that emerged. Though external, this difference had substantive, halakhic-legal, ramifications, as described above; it is thus important to stress its impact on both the theory and the practice of Jewish law.

Arnaldo Biscardi (Milan: 1972), vol. 2, pp. 532–536. (I am grateful to Prof. A.M. Rabello for this reference.)

32 See Schacht, *op. cit.* (note 29 above); Baron, *op. cit.* (note 30 above).

33 Elon, *Jewish Law*, vol. 3, pp. 1466–1467.

3. The Responsum as a Legal Opinion

i. Hearing One Side in the Other's Absence

The end of the Geonic period marks a new phase in the history of the responsum. Outstanding scholars can no longer impose their authority on outside communities, on the one hand, and no one halakhic institution can command acceptance of its authority by the various Diaspora communities, on the other. The independent power of each community and its chosen leaders gains ascendancy, as remarked on in a responsum by R. Solomon b. Adret (Rashba): "Individuals ... [are subject to] the men of their city as all Israel was subject to the Great Court or to the king."[34]

The fact that each community was independent, exercising authority over its members, led to tension between the local judicial institutions and the respondents. While conflicts involving outside respondents were not uncommon, there was often tension even between communal judicial institutions or elected leaders and respondents, who, as members of the local community, were under their jurisdiction. Responsa were sometimes perceived as unwanted interference in the affairs of the local halakhic authorities and institutions, though the competence, scholarship, character and erudition of the respondents were acknowledged by all concerned. These factors – the respondents' lack of formal status as legal authorities, and community dissatisfaction with their informal legal activity – brought about two changes in the status of the responsum, which we will now consider.

The first change, as noted above, was that the responsum was no longer regarded as a judicial ruling, but rather, as a learned legal opinion. To be sure, the form of the responsum was usually that of a court ruling, giving rise to the view, in many cases warranted, that this was indeed the intention of the respondent. A theoretical framework was needed to legitimate the practice of issuing responsa and clarify that the responsum was not what it appeared to be – a court ruling – but merely an opinion. In a number of responsa there are indications that the respondents were aware of the problem; we now turn to a closer examination of the factors that gave rise to it.

The most extensive discussion of the problem is found in a responsum by R. Solomon b. Simon Duran which sheds much light on the difficulties confronting the respondents. The responsum is as follows:

---

34 R. Solomon b. Adret, *Responsa*, part 3, #411.

> I have considered all that you have copied from the *Sefer Hamitzvot Hagadol*, section 208 of the negative injunctions, which states that a judge must not hear evidence from only one of the litigants, etc. This is known to us, and also attested to in bShevuot 31a and bSanhedrin 7b. We have not transgressed any injunction of the Sages. We are aware of their injunction, we have seen it, considered it, and held it in high esteem. But this injunction is irrelevant to the present issue because it applies only to a judge. But my late father R. Simon b. Tzemah Duran (Rashbatz) of blessed memory and other learned authorities who preceded him and his contemporaries, all agreed to respond to the queries of petitioners. They did not assert that the ruling in the case must be in accordance with their opinion. But they said, as it were, if the state of affairs is so and so, then the law is such and such. And it is up to the local judges to ascertain whether the facts of the case are as stated by the petitioner. If so, they rule according to the responsum, unless they find something to criticize, in which case they reexamine the matter until the correct decision is arrived at. The masterful judges of Israel, whose intention was to arrive at the correct decision, have acted thus.[35]

After stating his opinion, the Rashbash brings a number of arguments for it, which we will examine in turn.

> Rava said to R. Papa and to R. Huna b. R. Joshua: When one of my rulings is brought before you and you find something objectionable in it, do not tear it up before you come to me. If I have good reasons I will tell you, if not, I will revise my opinion (bBaba Batra 130b.) Surely he was referring to a decision issued as a responsum. Had Rava acted there as a judge, he would have had no reason to make such a statement. But surely they were judges and Rava was their consultant through a responsum which they used in their discussions with the other judges. This passage cannot be explained by saying that Rava had judged a case and given his ruling in writing, and then a similar case came before his students, who sought to use his ruling as a precedent in their own case, and he therefore commanded them as he did. Because he told them, do not tear it up, and if this refers to a past ruling, it is a fait accompli and they cannot reverse it. But it is certain that Rava refers to the future.

Most of the commentators accept the view rejected by R. Solomon b. Simon Duran. That is, they are of the opinion that Rava is referring to a court ruling he handed down.[36] The objection raised by R. Solomon b. Simon Duran,

---

35 R. Solomon b. Simon Duran, *Responsa*, #230. On this issue, rather than the related question considered below at note 58ff., see R. Jacob Molln (Maharil), *Responsa*, #193–195. And see responsa #76, #77, #172.

36 See, for example, the commentary of the Rashbam (R. Samuel b. Meir) on bBaba Batra 130b s.v. *psaka dedina:* "Decisions and arguments in a ruling I personally

that, once handed down, a ruling cannot be annulled, is not valid. It is quite conceivable that a document would be destroyed to prevent its future use as a precedent.[37] Furthermore, it is unlikely that Rava would use the term "ruling" to refer to a responsum, since, at least according to the Rashbash, his clear intention is to refer to an opinion that is not a court ruling.

The Rashbash's comment, "they were judges and Rava was their consultant through a responsum which they used in their discussions with the other judges," reflects the circumstances of his own time, but does not necessarily reflect the situation that obtained at the time of Rava and his eminent students R. Papa and R. Huna b. R. Joshua, who were senior judges and heads of academies in their own cities.

To bolster his position, R. Solomon b. Simon Duran advances yet another argument based on cases in the Talmud:

> And in the third chapter of bBaba Batra (41a), R. Cahana said to R. Judah who ruled according to R. Simon b. Eleazar: I am going to bring a letter from the

---

issued to someone, do not tear up and annul." Other commentators explain it similarly.

37 Shochetman (see note 8) uses this passage in an attempt to demonstrate that in the talmudic era the arguments for a ruling were not communicated. His reasoning is as follows:

> Firstly, Rava cautioned them not to tear up his rulings before asking him, so that he could give them his arguments. It seems, then, that the arguments upon which the decision was based were not included in the ruling itself (p. 327).

I think that it could well be the other way around – if reasons for the ruling are not given, but instead, these rulings resemble decisions for which no justification is cited (*psikata bealma*), such as those he cites on pp. 348ff., how can they possibly be critiqued? Shochetman himself acknowledges this (*ibid.*, note 24). His second argument is that

> Rava's request that his ruling not be torn up seems to prove that supporting arguments were not an integral part of the ruling, since had they been part of the ruling, their destruction would have been out of the question seeing that their contents would have included sacred halakhic discussion [which cannot be defiled by disposal].

But here the act of destruction referred to is not one of profanation of a religious text, but rather, it is a symbolic act on the part of the court to convey its reservations about the decision expressed in the document that is rended. Courts perform similar symbolic acts of tearing with notes or mortgages that have been paid.

> West [the land of Israel] saying that the law is not according to R. Simon b. Eleazar; in bShevuot 48b ... he said go and bring it. It is implied that they used to solicit responsa when they had queries about the rulings of their judges, and they would reply to them, and there is nothing wrong with this. And this does not violate the injunction against contempt for the court.

But these talmudic passages are not relevant to the matter at issue. In the cases cited by the Rashbash, the verdicts are based on a decision that in a particular debate among the Sages, the law will be in accordance with the views of a certain Sage. The responsum solicited is purely theoretical, for it is in no way related to the facts of the case being decided, but is an inquiry as to whether or not the law is in accordance with the opinion of R. Simon b. Eleazar.[38] From this passage, then, it is not possible to determine the procedures used to reach a verdict in cases turning on particular factual circumstances.

Furthermore, examination of examples brought by scholars[39] to establish the existence of a responsa literature as early as the talmudic period[40] reveals

---

38 And consider the following:

> Although it is forbidden, on the basis of the commandment against taking the name of the Lord in vain [Exod. 20:7], for the judge to hear only one side, and this must not be done, nevertheless this case is different, because it does not depend on the claims of the litigants, but on the laws of inheritance, which are established, and the testimony of witnesses makes no difference (R. Moses Isserles, *Responsa*, #112).
>
> It is well known from the responsa of all the Later Authorities that regarding a monetary dispute a responsum must not be issued to one of the parties. But to give a general opinion on matters pertaining to taxes, which are universally relevant, need not give rise to any apprehension that litigants might learn to lie from it. Further it is not a responsum to one side only but a general ruling for all communities (*Noda Biyehuda*, 1st ed., #50).

The attitude expressed by these respondents is similar to that of the Rashbash, but the analogy is incomplete, because these responsa make reference to particular circumstances of the given cases, circumstances described in posing the questions; here, then, specific rulings are to be handed down; thus these are not cases of rulings turning on classic halakhic controversies.

39 Mueller, J. *Briefe und Responsen in der Vorgeonischen Judischen Literatur* (Berlin: 1886); Elon, *Jewish Law*, vol. 3, pp. 1454–1455.

40 See R.J.N. Epstein, *Mavo Lenusah Hamishna* (Jerusalem-Tel Aviv: 1964), pp. 699ff. for examples. Epstein examines this early correspondence in the context

that almost all such responsa deal solely with queries about ritual law posed by scholars and heads of academies, not by members of the public. It also reveals that there is no proof that these queries were posed in order to resolve concrete disputes; rather, they seem to have been theoretical questions which arose in the wake of lectures at the academies or contemporary court rulings. The mere fact that questions were asked and responses were given in writing, or dispatched to distant locales, does not in itself have any bearing on our concerns here. It simply reveals the mechanics of the solution found to overcome the problem of the vast distances between the various centers of halakhic scholarship.

Nor can R. Solomon b. Simon Duran rely on the next talmudic source he cites:

> We also learned: R. Safra said in the name of R. Johanan: Two who are headstrong in litigation, one says, I want to litigate here, and the other says, let us go to the seat of the academy, he is compelled to try his case in his own town. If the need arises, they send a letter to the academy.[41]

This passage is irrelevant because a response to such a query by a court adjacent to the academy surely does not constitute a responsum as conceived by the Rashbash. A responsum of a Great Court is a definitive court ruling with which the court in the litigants' home town must comply; it is not at liberty to rule differently. Indeed, should they nevertheless choose to do so, the local judges would be considered "rebellious elders." But the words of the Rashbash do not preclude the possibility of the local judges' finding fault with the solicited response and deciding not to rule in accordance with its recommendations.

These sources indicate the existence of two systems, which, though probably constituted by the same individuals,[42] are distinguished from each other by their respective roles and powers. In one system scholars and legal experts

---

of discussing whether halakhic material was committed to writing in the period in question. In principle, of course, it makes no difference whether the responsum was written down or was transmitted orally via messengers who travelled between Babylonia and the land of Israel. See Z. Fraenkel, *Mavo Leyerushalmi* (Breslau: 1870), pp. 60ff., on *rabanan nehutai*.

41 bSanhedrin 31b, jSanhedrin 3:2, 21a.

42 See Y. Gafni, "Yeshiva Umetivta," *Zion* 43 (1978), p. 12; D. Goodblatt, "Hitpathuyot Hadashot Beheker Yeshivot Bavel," *Zion* 46 (1981), p. 14; also see the bibliography listed there in note 1.

correspond among themselves, while in the other, lower courts query and are guided by the Great Court. There is no evidence of courts querying scholars.

R. Solomon b. Simon Duran continues as follows:

> Had you taken note of the responsum of Maimonides, of blessed memory, to R. Pinhas the Judge, who lived in Neve Amun, it would be clear to you how individuals turned to him and how he responded, and who among us is greater than he was.[43] And this was the conduct of the greatest scholars, rabbis and sages. And one who ignores this is turning away from the cries of the needy, as it is said in the Talmud (bShabbat 58a): and one who responds observes the precept of "teaching the children of Israel."

This is R. Solomon b. Simon Duran's final argument. It gives the impression that one who refuses to answer petitioners is a transgressor, who "turns away from the cries of the needy," while one who responds fulfills an important religious precept. A closer look at his sources raises doubt as to the degree to which they actually support these assumptions.

The first passage cited by the Rashbash in this closing argument comes from bShabbat.

> R. Judah was sitting before Samuel. A woman appeared before them and raised a cry but Samuel took no notice of her. He asked: Do you not agree sir, that "One who stops his ears to the cries of the poor, he shall also cry himself and will not be answered?"[44] Samuel answered, Sharp one[45] that you are, your master remains cold, for you put the heat on the master of your master. For his excellency Mar Ukba, head of the court, sits in judgment, and it is written: "O house of David, thus saith the Lord: Execute justice in the morning, and deliver the spoiled out of the hand of the oppressor, lest My fury go forth like fire, and burn that none can quench it, because of the evil of your doings."[46, 47]

Rashi explains Samuel's answer Samuel to R. Judah as follows:

> I, your teacher, will not get burnt, but Mar Ukba, who is superior to both of us, and head of the court, will be burnt [will receive divine punishment]. As it is written, O house of David .... Only those who are in a position to judge are liable.[48]

43 See Maimonides, *Responsa* (Blau ed.), #355.
44 Prov. 21:13.
45 The nickname "sharp one" (*shinana*) is interpreted by some commentators as "a keen student"; the Geonic interpretation is "long-toothed." Cf. *Arukh* s.v. *shen* (4).
46 Jeremiah 21:12
47 bShabbat 55a
48 Rashi *ad loc* s.v. *resheikh bekriri*.

Apparently, far from supporting the view of R. Solomon b. Simon Duran, this passage contradicts it.[49] Samuel, who was the head of the academy – R. Judah was his student – **refused** to respond to the woman, on the grounds that the courts were the proper channel for her to use in seeking a remedy. If R. Solomon b. Simon Duran does not consider himself to be a court, and indeed, he states that he does not, then he does not incur the divine curse, but rather, the court does.

The second talmudic source is in bKeritot.[50] It suggests an exegesis of the verse enjoining the priests to "teach the children of Israel,"[51] namely, that there is an injunction against teaching when drunk. "To teach," states the *baraita*, is "to issue rulings," specifically, to rule on matters of ritual law (*isur veheter*).[52] But leaving aside the nature of the rulings in question, nothing is said there of this being an important religious precept, all the more so regarding a matter that is already *sub judice*. Furthermore, the talmudic context is explicitly that of actual rulings.

A review of these arguments leads – inescapably, it seems – to the conclusion that none support the view of R. Solomon b. Simon Duran. Yet this conclusion is problematic. How could he have understood these sources so differently from the other commentators? Indeed, this possibility cannot be seriously entertained. Rather, all the alleged corroborative evidence is no more than a rhetorical device intended to convince the petitioner of the correctness of the respondent's [i.e., the Rashbash's] position. As the Jerusalem Talmud puts it, "what is not explicitly stated, we support from many different sources."[53]

49 R. Menahem Hameiri comments: "It is true that the most eminent person is the best equipped to protest, and the most harshly punished. Nevertheless, those of lesser dignity are also punished since they could have lodged a protest but did not" (*Beit Habehira,* Lange ed. (Jerusalem: 1976), Shabbat, p. 203). It is possible that R. Solomon b. Simon Duran used this interpretation, according to which even one who has no absolute duty to issue an opinion is nevertheless liable if he failed to protest. But this interpretation is problematic. To suppose Samuel is at all liable for not heeding the woman's clamor raises the question of why he failed to answer her. For an interpretation along the lines of Hameiri and the Rashbash, see responsum #24 by R. Solomon Luria; and see the text at note 57.

50 bKeritot 13b.

51 Lev. 10:9–11.

52 Rashi *ad loc.* s.v. *hahoraa shel isur veheter*. It is stated in bShabbat *ad loc.* that the Amora Abba Arikha (Rav) did not teach on festivals, because he drank wine during the festivals, and his teaching was tantamount to his issuing legal rulings.

53 jBerakhot 2:3; 4c.

The only real argument in the responsum is precedent: the earlier authorities, even Maimonides, all engaged in writing responsa, an argument which R. Solomon b. Simon Duran includes along with the other purported demonstrations. If we keep in mind the conclusion of his responsum #529, quoted above,[54] in which he states that because of the ineptitude of judges and officials it is meritorious to respond to queries, even given the prohibition against acting the counsel's part, it is easier to explain the vindication of the institution of responsa proffered by the Rashbash in responsum #230, which we have just analyzed at length, an ironic and angry vindication based on shaky evidence.

What is the connection between the two responsa? In #230, the Rashbash establishes that responsa do not have the status of court rulings, and it is because – indeed, only because – this has been established, that the question of why, in issuing responsa, it is not the case that the respondents are "taking the part of counsel," could have arisen at all. This question is answered in responsum #529.

In one of his responsa, R. Solomon Luria adopts a view very similar to that of R. Solomon b. Simon Duran:

> I was asked... and at first it occurred to me not to reply at all, not even what a fox may dig out of a furrow [a triviality]. It is written [Deut. 32:7], Ask thy father, and he will declare unto thee, thine elders, and they will tell thee; and indeed, I follow their example: for who is greater in our generation than our master R. Israel Isserlein, who wrote that it is not his custom to respond on matters of monetary law, not even in the form of an abstract discussion, unless chosen by both parties and only after hearing all their arguments, which is not the present case... nevertheless I thought: widows and orphans have a covenant with the Almighty, so it is preferable to choose their cause and be their counsel. And it is not like "taking the name of the Lord in vain," or admitting testimony of one litigant in the absence of the other, which is forbidden in bSanhedrin 7b, because these prohibitions apply to judges only, as Rashi explains there. But in the present case, in which a judge has been chosen by both sides voluntarily – a judge who is none other than my kinsman and relative by marriage, the great champion R. Joseph Cohen, a true teacher, and I am like a student who discusses a point with his teacher to find succor for a widow – perhaps, after hearing the arguments of both sides, the master will like my exposition and affirm it. And even if it is right to respond, no one is forcing me to get so involved in the matter, since I have nothing to do with the ruling, for the case is being tried by an able and resourceful judge; why am I letting myself get mixed up in this troublesome affair? Nevertheless I say it is an important religious precept to take pains for the

---

54 See above, near note 25.

widow's benefit even if there is someone else in charge; and this is supported by what is said in bShabbat 55a.[55]

From here R. Solomon Luria proceeds to explain the passage cited, which is, of course, also discussed by R. Solomon b. Simon Duran.[56] He concludes from it that the woman who wailed before Samuel had asked him to help her in the case pending before Mar Ukba, the presiding judge, and Samuel refused because he thought such interference might offend Mar Ukba. This explains why Samuel admitted that he deserved some divine punishment – he had refrained from helping the woman, who was presumably a widow.[57]

R. Solomon Luria implies, as does R. Solomon b. Simon Duran, that ordinarily it is not difficult to justify interference in the affairs of a judge who does not rule according to the law. Here, however, because the judge held an official appointment, and was qualified for his role, a certain tact was called for; hence the need to seek some fact, such as the litigant's widowhood – for it is incumbent on all to help widows – that would justify intervention. R. Solomon Luria makes a clear distinction between a judge likely to "utter a false report" by listening to one of the parties in the absence of the other, and a respondent who merely engages in scholarly debate "like a student who discusses a point with his teacher," offering an opinion that the teacher may affirm or reject.

Paradoxically, it is the existence of a court system that makes it possible for the respondent to intervene without transgressing the injunction against "uttering a false report," because on this conception of the respondent's role, the respondent is not a judge, but only a teacher, or, in the words of R. Solomon Luria, only a "student who discusses a point with his teacher."

ii. Enactments Forbidding the Hearing of One Side in the Other's Absence

Thus far, our discussion has focused on whether the institution of responsa violates the injunction against hearing one of the parties in the other's

55 R. Solomon Luria, *Responsa*, #24.

56 See the text between notes 44–49, and note 49.

57 According to the interpretation of R. Solomon Luria, the reasons for meting out any punishment to Mar Ukba, who had not done anything reproachable, are unclear. It is quite understandable, however, that R. Solomon Luria could not compare the case from bShabbat 55a with the matter under discussion, because he could not attach any blame to the conduct of his father-in-law, who was appointed by both parties; he is thus silent on this point.

absence. This question is related to a broader problem, namely, the injunction which specifically prohibits the issuing of a responsum to only one side, and states that it is only permitted to issue a responsum to both sides, or to the presiding judges, at their request.[58]

In a responsum, R. Moses Isserles states:

> This is an enactment of the communities at the initiative of R. Haim of Vienna of blessed memory and other prominent people: Any rabbi who judges a lawsuit without hearing both parties – if, say, one comes to him and says: "Listen to me and write as you see fit," and he writes it for him – such a rabbi shall never judge any case again.[59]

A number of points should be noted. The prohibition is an **enactment,** therefore it does not obtain in biblical or talmudic law. A distinction is made between a "rabbi" and one who "judges a lawsuit." "Writing" the opinion for the petitioner is emphasized, while the injunction against hearing one of the parties is not contingent upon giving a written opinion. This implies that the reference here is to a particular case arising from special circumstances. Indeed, R. Moses Isserles states:

> A scholar must not write a ruling for one of the litigants in the "if so" format, or write him an opinion without a ruling, if he has not heard both sides, lest through his words the litigant will be learn to lie. Also because afterwards the other side will say something else and he will have to write a different opinion, and this leads to contempt for scholars.[60]

He distinguishes two kinds of unacceptable conduct: issuing **rulings** on "if so" cases, and rendering a written opinion without a ruling. In this he is in agreement with the view of R. Israel Isserlein, who writes:

> Concerning your question about [the estate of] the brothers and sisters, I am not answering at all, since it is a case of monetary law. And it has been the custom of

58 In this context, no distinction is made between written and verbal responsa; see Elon, *Jewish Law,* vol. 3, pp. 1501–1502; and cf. Maharil, *Responsa*, #193–195.

59 R. Moses Isserles, *Responsa*, #57; Also see *Binyamin Zeev* #156, #400; *Knesset Hagedola* (Jerusalem: 1970), Hoshen Mishpat 17:19.

60 R. Moses Isserles, *Commentary on the Shulhan Arukh* [Mapa], Hoshen Mishpat, 17:5.

> our forefathers of blessed memory not to answer, even if the matter is presented as an "if so" case.[61,62]

A similar enactment is mentioned by R. Moses of Padua:

> Recently my colleagues and I enacted a restriction which we took upon ourselves, not to answer disputes about matters of monetary law if not requested to do so by both parties or by the appointed judges. We saw that such conduct led to much corruption in our land, because each side took secret counsel surreptitiously, and fraudulently acquired a scholarly opinion to support its claim. This enabled the sides to come forth with "rulings" from learned rabbis and to establish claims to the said rulings, resulting in discord between rabbis and members of the community (*baalei batim*).[63, 64]

A ruling of the "if so" type is a ruling that begins with the words "if the facts are as described in the question," or some similar formula. It shows that

---

61 The copyist notes: "At another place he told his friends not to tarry." Note this comment regarding another responsum: "I wrote you this answer, as opposed to a ruling, in response to your question, because I did not hear the other side" (*Psakim Ukhtavim* #214); cf. #267.

62 *Psakim Ukhtavim* #62. His remarks are mentioned by R. Solomon Luria in the responsum discussed above (#24).

63 Compare another of R. Moses of Padua's responsa:

> You have written that I have been selected to be a judge and mediator… let me say that in no other way am I permitted to answer questions of monetary law, for I have an agreement with my colleagues not to answer questions on monetary law unless so requested by both parties (#44).

And cf. his responsa #45, #59, #60, and #66, and the discussion below. The *Knesset Hagedola* describes this as "a scruple of the pious" (Hoshen Mishpat 17:16).

64 R. Moses of Padua, *Responsa*, #40. *Divrei Haim*, part 2, Hoshen Mishpat #56, comments:

> It is proper not to answer questions of monetary law without hearing both parties, but I do not give a ruling to be carried out in practice, such as A is liable or B is exempt, I only advise in the form of a responsum. The law itself is set down by the great authorities of our times, and I only assess the petitioner's position. … and try to give him good advice, which is not at all like a ruling.

This misses the point of the arguments of both R. Solomon b. Simon Duran and the enactment.

the respondent has not carried out an examination of the factual evidence, but gives his ruling solely on the basis of the data supplied. Nevertheless, this does not mean that such a decision is not considered a ruling, since it is distinguished from "an opinion without a ruling."

It is extremely common for responsa of the Geonic period,[65] to open with the phrase "if so." This is understandable in view of what was said above. A Geonic responsum carried the weight of a court ruling. Since the Gaon did not hear both parties – in fact, he did not hear either party in a manner that provided him with a reliable impression – the validity of the ruling was qualified and made subject to the veracity of the data presented; no factual issues were determined by the respondent. Now this is a crucial point. The prohibition against hearing only one of the sides to a dispute apparently applies only to the determination of the facts of the matter. For determining the outcome of a particular case, however, it is not necessary for the court to hear the arguments of the sides, because it can be assumed that the court knows the law. Certainly, there is no need to fear that the court will be influenced by an argument made by one side in the absence of the other, whether or not the other is to have its say afterwards.

In principle, this method of avoiding the injunction against "taking the name of the Lord in vain" should have been acceptable to halakhic scholars in the post-Geonic period as well. However, they refused to adopt it, not for halakhic, but for practical reasons. Though they did not see writing an opinion as a transgression of the prohibition, they nevertheless opted to prohibit the writing of opinions unless both parties had been heard.

It is probable that they did so in order to avoid offending judges in the various communities, who were not, as a rule, halakhic scholars, but were considered laymen.[66] When R. Moses of Padua speaks of "discord between rabbis and members of the community,"[67] he is referring to these lay judges, often

65 Here are a few examples from the many available: *Kovetz Shaarei Tzedek*, 'gate' 4, sec. 38; part 4, 'gate' 6, sec. 8; 'gate' 8, sec. 11.

66 See S. Assaf, *Batei Hadin Vesidreihen Ahar Hatimat Hatalmud* (Jerusalem: 1924); M. Frank, *Kehilot Ashkenaz Ubatei Dineihem* (Tel Aviv: 1938); I. Baer, *Toldot Hayehudim Besfarad Hanotzrit* (Jerusalem: 1965); Elon, *Jewish Law*, vol. 1, pp. 20ff. On violent and ignorant judges in Spain, see Baer, *ibid.*, pp. 185, 206; on the same phenomenon in the Germanic countries, see A. Grossman, "Avaryanim Vealimim Beashkenaz Hakeduma," *Shenaton Hamishpat Haivri* 8 (1981), pp. 135, 147–148.

67 Responsum #40, quoted above.

wealthy community leaders.[68] Responsa issued by rabbis became a cause of discord between rabbis, whose authority was based on scholarship, and laymen, whose power was based on their status in the community. Learned responsa had the potential to undermine the authority of such lay judges. The various enactments on this issue express the balance which the rabbis hoped to foster by restraining themselves from issuing halakhic opinions. Criticism of cases in which such courts ruled counter to the law is reflected in the responsum of Rashi quoted above,[69] and that of R. Solomon b. Simon Duran,[70] who does not mince words and is outraged by this phenomenon; these documents thus provide us with insight into the real parameters of the problem.

The reason for the enactments and the custom of not issuing a responsum unless the respondent was asked to do so by both parties is thus clear. Under these circumstances, the local judge could not have any objection. The scholar did not encroach on the authority of the judge, since ordinarily individuals are free to choose any arbitrator they wish, a scholar being as good a candidate as any. However, if a scholar gave a ruling on a pending case, it was not unreasonable for the judge trying the case to take offense and regard it as interference with his authority. It is also clear that if a scholar was asked to issue a learned opinion by the judges themselves, he could do so without hesitation. Since the authority of lay judges was restricted to monetary law, it is not surprising that the responsa speak of an injunction against issuing an opinion to one of the sides only with respect to matters of monetary law.[71]

---

68 There is much material on the rulings of these *baalei batim*. See, for example, *Noda Biyehuda* (R. Ezekiel Landau); *Drushei Hatzalah*, 3:12–13. Also see Assaf, *op. cit.* (note 66), pp. 7, 86ff. This issue is connected with that of appealing the rulings of various rabbinical courts. See Assaf, *ibid.*, p. 74; *Mishpetei Uziel*, part 3, Hoshen Mishpat 1; A. Mechlowitz, "Sofiyut Hadin Bemishpat Haivri," *Dine Israel* 1 (1970), p. 3; Shochetman, *op. cit.*, pp. 355ff. When respondents are asked for an opinion on a court ruling, they regretfully decline with the same excuses that are given when asked for an opinion on a case before the courts. See R. Eliezer b. Nathan, *Responsa*, #114; Maharam of Rothenburg, Prague ed., sec. 410; *Hut Hashani* #18, and others. This is understandable, as there is no real difference between the two situations. And see J. Katz, *Halakha Vekabala* (Jerusalem: 1984), p. 240.

69 Rashi, Responsa #74; see above near note 21.

70 R. Solomon b. Simon Duran, *Responsa*, #529; see above near note 25.

71 R. Isaac b. Sheshet Perfet also explains the need to desist from answering one side only as stemming from the fear that

> One will present his arguments before the judge and get a ruling in his favor, but upon examination of the other litigant, the evidence does not

Thus the responsa literature can attest to the relationship between scholars and community officials and leaders. When there is a tendency on the part of scholars to refuse to answer questions on monetary law, and a need to apologize for and explain their actions, we can assume a certain ascendancy of lay officials in the community's power structure, and vice versa. It is also clear that in the Geonic period there could be but one state of affairs – that of the ascendancy of scholars, because judges were appointed and controlled by the Geonim.[72]

The nature and scope of responsa, it seems, are not determined by jurisprudential reasoning, but rather, by the complex network of pressures and interests to which the respondents are subject, and by their subjective understanding of their role. The responsum could be regarded as either a judicial ruling or merely an opinion, depending on how the respondent chose to define it.[73] His choice was influenced by a number of considerations, which changed with the circumstances. Had the status of the responsa literature been unequivocal, it is doubtful whether the various considerations mentioned by the respondents for and against rendering opinions would have been considered relevant at all. This insight, and the fact that practice in this regard had changed, makes it possible to resolve a query posed by the author of the *Meil Tzedaka*. He quotes a responsum from *Avodat Hagershuni*, which states: "If the petitioner has been examined by the respondent, and it is certain that his intentions are to know whether he is right or not, but he has no intention, heaven forbid, of learning how to testify falsely from the answer, it is permitted to reply."[74] The question is:

> I indeed wonder, on what grounds did they allow themselves to be so indulgent, since I found that this prohibition originates from the highest authorities, from

---

> stand up and the ruling must be reversed, which leads to contempt for scholars (*Responsa*, #5, #179; cf. note 24 above).

R. Moses Isserles concurs, and quotes this argument (above, near note 60). *Knesset Hagedola*, Hoshen Mishpat 17:19, suggests a wider interpretation: "Apparently, according to the arguments of R. Isaac b. Sheshet Perfet, the prohibition applies not only apply to monetary law, as suggested by R. Israel Isserlein in *Terumat Hadeshen*, but to ritual law as well, and this matter needs further consideration." However, it is unclear how this could be inferred from the Ribash, who, unlike other commentators, does not present his argument as a matter of general principle.

72 See Baron, *op. cit.*, (note 30), vol. 5, p. 20, and the sources cited in notes 19 and 20 there; S. Assaf, *Tekufat Hageonim Vesifruta* (Jerusalem: 1955), pp. 59ff.

73 This may be a solution to the difficulties in the responsum cited above where Rashi seems to have conflated two distinct prohibitions (#74, see note 20 above).

74 R. Gershon Ashkenazi, *Avodat Hagershuni* (Frankfurt: 1694), responsa #44, #47, #58.

> the very source of wisdom as spoken by the great Sages of the Mishnah, who taught, do not take the counsel's part; further, Rashi says in the name of R. Yehudai Gaon[75]. ... since this is an injunction of the great Sages, who would dare to make distinctions and allow it? Nevertheless, one can see venerable scholars who rely on their own discretion to issue responsa to those whom they do not suspect of falsehood. So, you can surely do the same because I can vouch for your having no interest in the matter.[76]

Halakhic scholars indeed found ways to permit the issuing of responsa under various circumstances if the reasons for so doing outweighed the prohibition. Previously, we saw R. Solomon Luria's rationalization for helping out a widow.[77] One of the responsa from *Avodat Hagershuni* cited above[78] concerns a case involving orphaned children.[79] When the problem was in a distant place, and the responsum urgently needed, respondents did not hesitate to answer, including the Ribash,[80] the Maharam of Padua[81] and R. Moses Di Trani.[82] Respondents also replied unhesitatingly if the petitioner was known by the respondent to be one who would not use the responsum to obstruct the judicial process.[83] R. Joseph Caro once responded for the following reason:

> Therefore concerning this matter, where the question came from one side only, I refused to answer. Afterwards I saw that one of the scholars of this city had written a responsum on this question,[84] which I found wanting. I saw that if I kept silent it would be thought that I concurred. Therefore I said: As I was rewarded for withdrawing from this matter, so I will also be rewarded for examining it, thus I have been compelled to depart from my regular behavior.[85]

---

75 See the text above at notes 11 and 12.

76 R. Jonah Liedshoffer, *Meil Tzedaka* (Prague: 1757), #53.

77 R. Solomon Luria, *Responsa*, #24; see text above at note 55.

78 *Avodat Hagershuni* #44.

79 Cf. Rashi, *Responsa*, #74 (above at note 21), which opens with greetings to the petitioner, "Isaac the Orphan."

80 R. Isaac b. Sheshet Perfet, *Responsa*, #5.

81 R. Moses of Padua, *Responsa*, #24.

82 R. Moses Di Trani, *Responsa*, part 3, #21. See also *Sefer Beer Sheva*, p. 115; R. Moses Isserles, *Responsa*, #17, #57, #112.

83 Asheri, *Responsa*, #17; R. Jacob Molln, *Responsa*, #70; R. Tzvi Ashkenazi, *Hakham Tzvi*, #117; *Avodat Hagershuni* and *Meil Tzedaka*, mentioned in notes 74 and 76 above, and see note 63 above.

84 This refers to R. Moses Di Trani, see his responsum #80.

85 *Avkat Rokhel* (Izmir: 1797), #81.

Similarly, respondents answered in order to carry out important precepts, support worthy causes, and rectify wrongs.[86] The author of *Knesset Hagedola* states that

> In our times it is the custom that if a learned man is asked about a case, even by only one of the parties concerned, he writes a responsum in reply, apparently because the respondent assumes that the case itself will not be brought before him for trial and that therefore there is no need to fear thwarting justice. Nevertheless, from the opinion of R. Isaac b. Sheshet, it appears that this too is prohibited.[87]

It seems that the custom spread due to both the erosion of the prohibition in practice, and the increasing irrelevance of its underlying rationales. Circumstances had changed, the status of scholars had risen, and when they were unable to contain their desire to answer questions of law that had bearing on the petitioners' lives, or saw no important considerations militating against a reply, they did not withhold their answers.

iii. False Representation

Responsa have two dimensions: the temporal, aimed at the particular event which elicited the legal question to be decided, and the supra-temporal, which transcends the facts of the given case and is directed toward the future, toward deciding the law in a manner that establishes its position with respect to possible future cases.[88] The more pronounced the first dimension, the more

---

86 For example, consider the following passage from a responsum in *Avodat Hagershuni*:

> Certainly the learned R. Moses of Padua writes in #60 as follows, "Indeed, in a case in which the embellishment of a religious precept is concerned, our Sages took no heed of this apprehension." He treats the subject fully and brings arguments for his views. I regard the present case, too, as a matter of the embellishment of a precept ... (#124).

Responsum #53 of *Meil Tzedaka* concludes that in every case where such a pretext was found, the respondents answered.

87 *Knesset Hagedola*, Hoshen Mishpat 17:19.

88 In his "Towards a Jurisprudential Analysis of the *Kim Li* Argument" (Hebrew), Hanina Ben-Menahem cites a parallel distinction: "A judicial question is a question regarding the law which comes before a judge who must decide on a solution

forceful the argument against issuing a responsum to only one of the sides; conversely, ascendancy of the second dimension brings about a weakening of the argument. Thus we found scholars who issued their opinions as if taking part in an academic discussion, as "students engaged in discussion with their teacher." When issuing responsa was forbidden in this form as well, due to interference in the administration of justice that arose from rulings issued in the "if so" format, yet another form evolved. It is described by R. Moses of Padua:

> I attended to the following question about three brothers, Reuben, Simon and Levi. ... and behold, even though it was difficult for me to answer the quarrelsome arguments about money without hearing the brothers, though the petitioner assured me that it was a theoretical question of law that he wanted to learn, and stated the question in the form of "Reuben and Simon who did such-and-such" – for our venerable Sages used to answer questions asked in this manner – but since I was not unaware of the trickery of the litigants who tried to conceal their intentions and deceive me by framing their query in terms of a hypothetical question about Reuben and Simon ... .[89, 90]

Apparently, the petitioner, claiming that his sole interest was to learn the law, pretended to ask a purely theoretical question which purportedly had no bearing whatever on any actual situation. He presented the question in the form of "two individuals, A and B," that is, two hypothetical individuals,[91] but failed to fool the respondent, who detected this deception. R. Isaac b. Sheshet, however, was deceived by trickery on the part of a petitioner, in a case he writes about as follows:

> We have already corresponded with R. Makhluf b. Hanin about this conflict, and had I known that he was involved in this matter I would not have answered him at all, because one who gives a decision in a litigation and rules on a case before listening to the arguments of both sides together, is senseless and shameful. ... to be sure, I understood from his question that he left no stone unturned to have the beneficiary win the case, and he brought numerous arguments, and suppressed the arguments of the other side, and I said this at the conclusion of my responsum,[92]

---

to the dispute. ... a legal question takes the form: 'What is the position of the law on question X?'" (*Shenaton Hamishpat Haivri* 6–7 (1979–1980), p. 53). And see J. Raz, *The Concept of a Legal System* (Oxford: 1980), p. 218. I thank Dr. Ben-Menahem for drawing these sources to my attention.

89 See note 86.

90 R. Moses of Padua, *Responsa*, #60. Cf. #54.

91 Cf. Elon, *Jewish Law*, vol. 3, p. 1512.

92 R. Isaac b. Sheshet Perfet, *Responsa*, #145.

and I thought that he favored the beneficiary and hated the merchant, but it did not occur to me that he himself was the beneficiary, and a party to the case, and I gave my opinion according to the question as he posed it.[93] But now, as your question arrived, I see that there are great differences between your question and his. ... therefore I think proper to send you a copy of the answer which I sent him, so that you can see with your own eyes the differences between his arguments and yours, and since I responded to him according to his description of the case, I will also send you an answer according to your description.[94]

4. Responsa or Codes – Which Are Definitive?

We spoke above of the dual nature of responsa. On the one hand, they are pragmatic solutions to concrete problems, on the other, they have a trans-temporal relevance. Emphasis on the latter dimension makes it possible for those who issue rulings to rule on the basis of the responsa literature. On the other hand, the pragmatic dimension is pivotal in resolving the following dilemma: if a ruling as formulated in the various law codes conflicts with the ruling as articulated in responsa, which is to be preferred?[95] This question can

---

93 See text above at note 65.

94 R. Isaac b. Sheshet Perfet, *Responsa*, #5. See notes 24 and 63 above. Also see M.A. Friedman, "Shtei Teshuvot Min Hageniza," *Dine Israel* 9 (1978–79), pp. 165, 177. Cf. *Noda Biyehuda*, 2nd ed., Even Haezer #114:

> I have had a similar case, they came with their questions from a great distance, about an issue which seemed to me so simple, that I wondered why they had to come such a great distance with their question; and afterwards the other litigant came and told me that the facts of the case were quite different, and I said that I would wait for some time. ... but now I have information from a number of additional sources, and the other side is silent, so I answer.

See the end of this responsum; and *ibid.*, Hoshen Mishpat #49.

95 See I.Z. Kahana, "Hapsak Vehateshuva," *Mehkarim Besifrut Hasheeilot Vehateshuvot* (Jerusalem: 1973), p. 97; Elon, *Jewish Law*, vol. 3, p. 1456. Kahana did not gather all the relevant material on this issue; see *Sdei Hemed*, Klalim, Haposkim, sec. 5, which discusses contradictions between Maimonides' *Code* and his responsa. Cf. M. Benayahu, "Teshuvot Sheeilot Larambam," *Sefer Hazikaron Larav Nisim*, Hikrei Talmud, p. 175. Also see the Tashbetz, *Responsa*, part 1, #1, #151; the Rashbash, *Responsa*, #11, and *Sdei Hemed*, sec. 10, on contradictions between the responsa of R. Solomon b. Adret and his rulings; see R. Joseph Colon, *Responsa*, #69; see *Sdei Hemed*, *ibid.*, sec. 13 on R. Joseph Caro, sec. 14 on R. Moses Isserles, sec. 15 on R. Israel Isserlein, the Maharil, the Rashbatz, R. Haim Joseph David Azoulay and others, and other works.

be approached from two different directions. One criterion is the date of authorship of the works in question, the simple but not always decisive assumption being that the later of the two rulings is decisive.[96]

While this criterion can be useful in resolving inconsistency in the works of a single author, it is useless in resolving contradictions between two authorities, in which case other criteria for determining which ruling takes precedence must be used. The second criterion works by evaluating both the responsa and the codes from the trans-temporal perspective. The relative substantive value of the contradictory rulings is examined. When the matter at issue involves ritual law, an area in which the authority of respondents to render opinions has never been questioned, clearly, no conflict arises between the pragmatic and the trans-temporal outlooks. The problem arises only when both concerns, the pragmatic and the long-term, come into play in the context of responsa involving matters of monetary law. Let us briefly examine a number of examples.

Some scholars who consider this question advance the following argument: "it is possible that the author of a code wrote theoretically – "as a matter of law" – but did not elaborate the law as applied "as a matter of actual practice," while in his responsa he took much more care to set out his reasoning precisely in order to respond to the petitioner in a practical manner."[97] The expressions "as a matter of law" (*lehalakha*) and "as a matter of actual practice" (*lemaase*) refer to what we have called the trans-temporal, ideal dimension, and the time-bound, pragmatic dimension, respectively. It is not

---

96 See R. Joseph Di Trani, *Responsa*, part 1, Hoshen Mishpat #7, and the criticism in *Veyosef Avraham*, p. 182. On the dating of responsa, see Elon, *Jewish Law*, vol. 3, pp. 1458–1459, note 19.

97 See R. Moses Benvenisti, *Pnei Moshe*, part 1, 12, s.v. *vehinei* (27b); and cf. the remarks of R.A.Z. Margalioth, cited in Elon, *Jewish Law*, vol. 3, p. 1458, note 18:

> Responsa are more reliable than abstract rulings, because the decisions in responsa are intended to be carried out, thus the respondent takes more care than those who write theoretical decisions for themselves, which may be intended only as academic discussions not meant to be put into practice (R.A.Z. Margalioth, Introduction to *Shaarei Teshuva*, Orah Haim; our translation).

If we take him literally, he refers to those who write theoretical decisions "for themselves," which is quite different from writing legal codes for publication so that the rulings therein can guide the readers' actual conduct. And see note 99 below.

argued that the legal authority (*posek*) in every case intended only to establish ideal norms; however, it is suggested as a possibility that must be kept in mind. With regard to a responsum, on the other hand, the probability of a decision's being intended solely as a "law of aspiration" is so low that it can be safely discounted. However, this argument is open to criticism.

First, the argument regarding the alleged idealized or abstract nature of the codes cannot be generalized to apply to all the codes, since these works are highly diverse in nature. The codes of Alfasi and Asheri are different from the *Arbaa Turim*, the *Beit Yosef*, and the *Shulhan Arukh*.[98] Indeed, it seems the argument was actually made only with regard to contradictions between Asheri's code and his responsa. This argument applies only to the kind of codes represented by that of Asheri, and no conclusions true of codes and responsa in general can be drawn from it.[99] It can hardly be suggested that works addressed to the general public, and written with the intention of defining lawful conduct, are not intended to be put into practice.[100]

Second, in many responsa the authors are careful to qualify their decisions with the statement that they decide the questions from the theoretical point of view only, and that their rulings are not to be put into practice.[101]

---

98 For more on this see Elon, *Jewish Law*, vol. 3, pp. 1138ff.

99 Margalioth may have been making a similar point, see note 97 above.

100 In his introduction to *Torat Habayit*, the Rashba writes:

> I will write ... using a dialectical style ... and afterwards I will write brief legal rulings ... and the reader will fulfill his duty to study by evaluating this work with careful diligent research so that he can apply the rulings, and in the end he will be able to conduct himself according to the short rulings.

Similarly, we find the following in Caro's introduction to the *Shulhan Arukh*:

> I thought it good to gather the sayings, which are as beautiful as flowers and jewels, in a short form ... in order that everyone in Israel will be fluent in the law of the Almighty, and if a scholar is questioned on a point of law, he will not stammer ... every law which might be asked will be clear to him, so that it can be carried out in practice. ... furthermore, young pupils can read it constantly, and learn it by heart, and what they have absorbed at a tender age will be readily recalled and carried out in practice from childhood on .... they will amuse their souls by studying this work, which is full of delights, it is decided law (*halakha psuka*), requiring neither explanation nor debate.

101 See Elon, *Digest* (see note 1 above), p. 13, the text at note 28 and note 28; *Jewish Law*, vol. 3, p. 1473, and note 66.

Third, if we seriously entertain the view that responsa on monetary law not requested by both parties or by judges hearing the cases were intended as learned opinion only, that the respondents had no intention of deciding them, but sought merely to present their ideas to their teachers, the judges; the view that they are merely theoretical essays not meant for implementation in practice, then, even if this is not outright subterfuge, it cannot be reconciled with the theory that in case of conflict responsa take precedence over codes. To be sure, this objection applies only to a limited sphere – to responsa on monetary law, and to respondents who take the position that their responsa are no more than theoretical opinions. But granted this, it is nonetheless clear that it is necessary to further examine the claim that responsa should be preferred to rulings found in codes.[102]

Let us examine a famous responsum on this matter, from R.N.Z. Berlin of Volozhin to the sons of R.Y. Bamberger, who were in doubt about whether to

---

102 A distinction must be made here between the maxims "act on the basis of laws intended for implementation" and "acts are authoritative." The first apparently applies both to halakhic rulings and to actions regarded as precedent-setting. (For the status of such actions, see Elon, *Jewish Law*, vol. 2, pp. 945–946.) Consider the following *baraita*:

> The Sages taught: The law must not be established from study or from an action, until it is clearly stated that the law is to be put into practice (bBaba Batra 130b).

We cannot explore all the interpretations suggested for this passage by the various commentaries (Rashbam, Yad Rama, R. Yomtov of Seville) here. Apparently, this *baraita* warns against making hasty generalizations about the law, either on the basis of a teacher's assertion in the course of a lecture that the law is decided in the manner he has just explained, or on the basis of an action performed by the teacher. It must be stated explicitly or deduced with certainty (see bBerakhot 24a, bShabbat 54a; bBeitza 29a) that the intention of the teacher was to teach the law "as it is to be put into practice."

This *baraita* thus does not entail any precedence of learning over action. The most far-reaching interpretation of it is that of the Ritba, who says that the action of a teacher with respect to a given matter should be regarded as an instruction on how the law is to be carried out in practice, unless the teacher states that it is an exigency ruling only.

The principle "acts are authoritative" (see variant readings in bBaba Batra 83a) definitely points to the precedence of action over "law intended for implementation" (see bShabbat 126b; and cf. bKeritot 13b). Furthermore, this is so not only regarding the actual performance of an act, but even the instruction that others are to perform a certain act – see bBaba Batra 82b, 83a. The

print their father's responsa, because he had made the request that his responsa, "which he answered on matters to be put into practice," not be printed, "because responsa are less reliable than works written on the subject which treat the issue more comprehensively than one who answers when asked"[103]; he had concluded by quoting the Talmud,"One must not learn the law from practice."[104, 105]

R.N.Z. Berlin dispels the apprehension of the inquirers:

> In my humble opinion this opinion is not well founded. On the contrary, before answering a specific question to be put into practice, one studies the law more profoundly than in the course of ordinary studies,[106] and there is more divine assistance when dealing with a practical matter.[107]

It is clear that divine assistance cannot be demonstrated. Though there is a reference to divine assistance coming to the aid of the halakhic scholar to facilitate his ruling's being accepted in a controversy,[108] we found no reference at all to support the claim that the aid so bestowed is more plentiful for one dealing with a practical question. Lack of ability to find this reference might

---

Maharil's well-known responsum #76 (Tacz ed. #72 (3)) must be regarded in this light. I refer to the following passage:

> You have written that one must not rely on responsa. The contrary is true – they are practical rulings, and we can learn more from them than from codes because the authors of the codes were not present when the cases were decided, as is written in the Talmud... [here he cites bBaba Batra 82b] and the rulings handed down in responsa on the cases are all intended to be put into practice.

103 Compare the following:

> Because these are words written in greater deliberation with regard to the roots of the law....because sometimes one cannot go into the matter deeply enough due to the urgency of having to give an answer" (R. Jonah Navon, *Nehpa Bekesef* (Constantinople: 1648) part 2, Even Haezer #27).

And see below.

104 See note 102 above.

105 See the editor's foreword to *Responsa Yad Halevi* (Jerusalem: 1969).

106 See note 103.

107 *Meshiv Davar* #24.

108 "The Almighty is with him – the law is in accordance with his view in every instance" (bSanhedrin 93b). Also see *ibid.*, 106b, to the effect that even one of

not be conclusive proof to the contrary, but let us examine the reference R.N.Z. Berlin himself brings to support his position: "It was stated by the Sages in bKetubot 60b that divine assistance is given in profusion to bring forth a correct halakhic decision."

In the talmudic passage he cites, Abbaye issued a practical ruling, but learned that he had erred in so doing when a quotation from Rab and Samuel was drawn to his attention, and he was forced to retract his ruling.

> Abbaye said, This bears out what the Sages said, that one must not give a ruling even on the permissibility of mixing eggs with porridge, when his teacher is present, not because it is disrespectful, but because he does not get divine assistance in what he is saying. For I am familiar with this saying of Rab and Samuel, yet my decision was not helped by Heaven.

In other words, though Abbaye knew the law, he did not rule according to the law because he handed down his decision in the presence of his teacher R. Joseph,[109] which was unhelpful. To issue a ruling in the presence of one's teacher may indeed lead to withdrawal of divine assistance in making a proper decision. However, we do not know the extent of the role played by divine assistance in the ordinary course of giving a halakhic decision.

The second passage quoted by R.N.Z. Berlin is from a discussion in the Talmud quoted above: "until it is clearly stated that the law is to be put into practice."[110, 111] We have discussed this passage above.[112] His final proof is as follows:

> Also in bSanhedrin 86b, concerning a rebellious elder, who taught again as he had taught before, he is exempt. But if he taught that his ruling should be applied

---

great wisdom and knowledge of the law is not guaranteed that his ruling will accord with the law; and see below. On divine intervention in determining the law, see Elon, *Jewish Law*, vol. 1, pp. 240ff.; Y. Englard, "Tanuro Shel Akhnai–Peirusheha Shel Agada," *Shenaton Hamishpat Haivri* 1 (1973), p. 45.

109 See *Tosafot ad loc.* s.v. *ki afkiruta*; and *Kesef Mishne*, Laws Concerning Study of the Torah 2:5.

110 R.N.Z. Berlin relies on R. Samuel b. Meir (Rashbam) *ad loc.*, therefore it is only according to the Rashbam's interpretation that action can be saïd to be a preferred source of law; cf. note 103.

111 bBaba Batra 130b.

112 See note 102.

in practice – he is liable. It is clear that a ruling concerning an action carries more weight.

It seems to me that the reason for differentiating between the teachings and rulings of the rebellious elder is not because the law in a ruling is more deliberate, righter or truer than that expressed during academic discussion. The explanation has to do with recognition of the fact that uniformity of conduct is a desideratum to be achieved even by coercion, while in the academic context a diversity of opinion is tolerable.[113] Opinions of individual scholars, or the minority view, are not perceived as "less correct" than the majority view. They are recorded for future reference, so that a future court will be able to cite them and rely on them in its rulings.[114] Laws are decided in particular ways not because they are correct, but because they reflect the majority view.[115] Furthermore, judicial decisions must be accepted even if demonstrably wrong – even if the judges rule that left is right and right is left.[116] This also applies to rulings handed down in particular cases. The duty to comply with such rulings[117] is due not to their absolute correctness, but to their validity in having been issued by the proper authority. This validity is limited to

113 See Elon, *Jewish Law*, vol. 3, pp. 1061ff. and vol. 1, p. 288, note 15.

114 Mishnah Eduyot 1:5.

115 And see note 108 above.

116 Sifre Deut., 154 (Finkelstein ed.), p. 207, and see variants given there. In any case, the distinction between "academic" and "practical" teachings with regard to a rebellious elder is identical to that made with respect to the ruling of a student in the presence of his teacher (see text above at note 109 and the sources indicated in the note). This is also the distinction adapted by those respondents who refused to answer questions on monetary law if requested to do so by one side only:

> As a matter of fact, I reply only out of respect, for I do not want to refuse an eminent person, but only as if I were writing a legal essay, for I do not wish to be, heaven forbid, the lawyer's counsel… but like a student before his teachers, I shall discuss and explain all that is instructed from heaven, and I do not say to anyone, accept my opinion (Rema of Pano, responsum #89).

In all these cases the distinction is a normative one, and the status of the ruling, whether avowedly "academic" or "practical," depends on the respondent's intentions.

117 Unless the ruling misconstrued a Mishnah, which borders on the court's exceeding its jurisdiction.

the specific instance in time to which the ruling applies; if the ruling is erroneous, it makes no contribution to supra-temporal law.

Furthermore, it is not the reasoning, but rather, the conclusion – the instruction regarding how one must act – which carries the decisive weight.[118] This final instruction can be correct even if the respondent's reasoning, in whole or in part, is faulty. The complexity of the kinds of situation upon which judicial decisions usually rest generally generates a number of questions to be decided; a wrong answer to one of these questions does not necessarily cause an error in the final conclusion. Such a case is recorded in the Talmud.

> There was a group of herb-dealers. Each one of them baked for all. One day they told one of them: Go bake for us. He told them: You take care of my coat. While [he was] away, the others did not look after the coat and it was stolen. They came before R. Papa and he made them liable. The other Sages asked him, Why is that? It was neglected while they worked together with the owner. He [R. Papa] was silent. In the end it turned out that at the time the owner was drinking beer.[119]

It is related of one of the prominent halakhic figures of the past generation,[120] R. Haim Soloveitchik, that he avoided issuing halakhic rulings, letting

118 "When the judges reached a conclusion, the litigants were brought in, and the head judge declared: So-and-so you are exempt, so-and-so you are liable" (Mishnah Sanhedrin 3:2). The judge has no obligation to offer any reasons for his decision. See the thorough discussion in Shochetman, *op. cit.* (note 8). Even when the litigant is entitled to request that the judges show the source of their conclusion, they are not bound to cite their reasons, but only the arguments of the sides and the final ruling. See the references cited in the *Encyclopedia Talmudit* s.v. *dinei mamonot*, notes 553–555; and note 37 above. It should be pointed out that while most responsa give reasons for the decision, it is likely that there were at one time many responsa that gave none at all, and because of this, they were not preserved. Neither the respondents, nor the petitioners or copyists, had any interest in preserving them. A responsum by Maimonides (#14) consisting of only one word is an exception, which may prove the rule. Responsa of this type are similar to the rulings discussed by Shochetman, and, as he notes, both rulings and responsa were issued with and without reasons; indeed, as is the case with regard to responsa, very few rulings that lack reasons are extant.

119 bBaba Metzia 81a, cf. 81b; bGitin 89b; bAvoda Zara 22a.

120 M. Berlin, *Fun Volozhin bis Yerusholaim* (Yiddish) (New York: 1933), pp. 319–320. Also see M. Elon, "Od Leinyan Mehkaro Shel Hamishpat Haivri," *Mishpatim* 8 (1977), p. 119, note 52.

the local judge make all decisions. Once someone disagreed with the judge's decision on a certain matter. The litigants went to R. Haim and it was agreed that they would send the judge's decision, and R. Haim's decision, to the rabbi of Warsaw, R. Isaac Elhanan Spektor, who in turn would let them know by telegraph which of the two was right. R. Haim asked the rabbi of Warsaw to issue his ruling without specifying any reasons, "in order not to trouble him too much," as he explained in a letter. But those close to R. Haim were aware that there was another reason as well: had the ruling from Warsaw been handed down together with the reasons behind it, R. Haim might have found fault with the reasoning, but if only the decision arrived, he would put his trust in it, for it was handed down by R. Isaac Elhanan, whose authority he recognized.

It is told of R. Raphael Shapiro and R. Isaac Elhanan, that even when the latter brought an argument which the former found unacceptable, he nevertheless ordered that the R. Isaac Elhanan's ruling be followed, on the assumption that a prominent scholar (*haver*) would not issue a faulty decision, and thus, no mishap would result from his ruling. He reasoned that even though the arguments which R. Isaac Elhanan proposed did not constitute conclusive proof, it was likely that a more acceptable argument could be found. In other words, in a responsum, the rationale does not carry decisive weight; rather, the ruling is its essence.

So if a respondent enjoys divine assistance, this is aimed, as the examples indicate, at the practical instructions the ruling entails: the decision that something is forbidden or allowed, liable or exempt. The important thing is that no mishap result from the respondent's advice, and indeed, no mishap can result if these pragmatic conclusions are correct. In this sense, the connection between the reasons behind a ruling, which are trans-temporal in nature, and the ruling itself, which is bound in space and time to the specific case under consideration, is severed.

Furthermore, the responsa we have examined here are intended to decide concrete cases. The configuration of special circumstances of each case may well be unique, so that the relevant law, as cited by the law codes, the general law which would be applied by a scholar to determine the outcome of the case, might prove inadequate to the particular case at hand, and its application might yield a different decision than that arrived at by the respondent who has been sent specific information on the details of the case. This sensitivity on the part of the responsum, as opposed to a ruling in a code of law, makes it possible to tailor its decision to the specific parameters defining the context of the case. The decision is directly connected to this set of facts, and by virtue of this correlation becomes particularized in space and time – valid for this instance only. It follows that the responsum is not comparable to the supra-temporal legal ruling, which, as we stated, is intended to be eternally valid.[121]

Thus we can find no support in the sources for the view that rulings expressed in responsa take precedence over those found in codes; nor do we find any substantive arguments, that is, arguments based on characteristics of responsa and codes, to justify such a bias. The argument of R.N.Z. Berlin should be understood in the context of the petitioners' desire to avoid publishing the responsa in any form. The radical nature of their attitude evoked a similarly extreme reaction on the part of the respondent, leading him to exaggerate the importance of responsa. Thus this responsum cannot resolve the problem posed above: when a responsum conflicts with a judicial ruling, which is considered definitive?[122]

5. Conclusion

Examination of the responsa literature reveals that it is characterized by two contradictory perspectives. From one, it is oriented toward changing circumstances, while from the second, a different attitude emerges. Taken globally, the responsa appear to constitute a system of judicial rulings issued to litigants upon request. A closer look, however, reveals a clear distinction between responsa on monetary law and those that treat other branches of the law. If we look only at the former, a difference emerges between the Geonic period, during which a responsum was regarded strictly as a judicial ruling, and the post-Geonic period, during which this attitude relaxed considerably. Backed by their acknowledged official authority, the Geonim, as well as the halakhic authorities who preceded them, regarded their responsa as judicial rulings. They avoided the problem of issuing rulings without hearing both sides, as mandated by law, by qualifying rulings with the proviso that the rulings stood only if the facts as presented by the petitioner were correct. This qualification was articulated as the clause "if the facts are as submitted in the

121 To be sure, a legal ruling might be subject to interpretation, and in this sense also limited to particular circumstances; but this task must be carried out by the respondent, who must adapt the ruling to the case before him.

122 This also serves as an example of the fact that no sweeping conclusions should be drawn from a responsum on a particular matter.

question." Indeed, not only did they regard their decisions as full-fledged rulings, they regarded respondents who construed their decisions solely as learned opinions as having transgressed the injunction against "taking the counsel's part."

With the weakening of the centralized legal structure, and the dispersal and fragmentation of authority, the status of responsa as judicial rulings came into question. The basis for this challenge was not the law itself, for, as we saw, such a challenge could easily have been refuted, and a suitable legal solution to any arguments against the injunction found. Rather, the calls for relaxing the prohibition arose from the fragmented social structure of the Jewish communities. The splintering of authority, however, was only an indirect cause of the responsum's changing status. It, in turn, gave rise to a situation where local judges, the majority of whom were unlearned, could not tolerate outside interference, via responsa, in their decisions and the consequent damage to their status as judges, as deciders of the law. The scholars withdrew in two phases. First, they declared that responsa were only to be regarded as learned opinions, on the assumption that no one would argue with their expressing their opinions, which, after all, represented the Torah – the law. It was easier for them to rebut the charge that in doing so they were "taking the counsel's part" than to stand up to the demand that they refrain from maintaining what was in essence a parallel judicial authority.

Evidence of the next stage in the evolution of the role of the responsum comes from a limited number of locales, not from the entire Diaspora. This second stage is reflected in enactments forbidding the issuing of responsa to only one of the sides to a dispute, and restricting answers to queries posed by both sides or by the adjudicating judges. The technical argument, that the responsum was not a ruling but only an opinion, no longer sufficed to enable scholars to present their opinions. Because the legal argument was inadequate, it became necessary to resort to more practical measures. This led to internal divisions within the responsa literature: the status of a responsum was determined by its subject matter; and with respect to responsa on matters of monetary law, there developed a further differentiation between responsa intended for actual implementation and those that were theoretical in nature.

Yet another source of differential status was the essential duality of the responsum, which, on the one hand, aims at resolving a particular concrete case, and on the other seeks to establish the law in a manner transcending the confines of its immediate context. This duality was exploited in attempts to justify endowing the responsum with a particular status: whichever aspect – the time-bound or the transcendent – was more convenient was emphasized at the expense of the other, both in determining the status of the responsum as opinion or definitive ruling, and in determining which type of decision took precedence, the responsum or the ruling. The various arguments are less than convincing, or at any rate, are at best equally convincing. And indeed, there is a reasonable explanation for this: in the final analysis, as we said, the responsum has a dual mandate, and attempts to ignore either dimension are invariably prompted by interests and circumstances not inherent in the responsum itself.

CHAPTER 4

THE LAW IS IN ACCORD WITH THE LATER AUTHORITY – *HILKHATA KEBATRAI*: HISTORICAL OBSERVATIONS ON A LEGAL RULE

Israel Ta-Shma

1. The Ashkenazic Understanding of the Rule
2. The Rules "the Law Is Like the Later" and "The Law Is Not According to a Student in the Presence of his Teacher"
3. "Jephthah in his Generation Is Like Samuel in His"
4. The Turning Point in the Middle Ages
5. Conclusion
6. Postscript (1994)

1. The Ashkenazic Understanding of the Rule

> This great veneration for the views of earlier authorities did not prevent Jewish law from establishing, over a period of time, a fundamental principle for determining the law – a principle that, on the surface, seems contrary to the principle of priority of the earlier over the latter, but which was essential in order to empower the authorities of later generations to make legal rulings responsive to contemporary problems and consonant with contemporary conditions. This important principle is that "the law is in accordance with the views of the later authorities" (*hilkheta ke-vatra'ei*), and it has existed since the geonic period.[1]

This is M. Elon's description of the substance and role of the rule "the law is in accord with the later authority" (henceforth, "the law is like the later"). After some discussion, he concludes:

> Thus was established and accepted the fundamental principle of decision making in Jewish law – "The law is in accordance with the views of the later authorities."

1 M. Elon, *Jewish Law* (Philadelphia and Jerusalem: 1994), vol. 1, p. 268.

> It should not be thought that this principle diminished in any way the respect that later generations accorded to the earlier generations. It was precisely this respect that induced the later authority responsible for declaring the law to ponder his own decision earnestly, fearfully, and humbly, because he was aware that he was dealing with a question already considered by the earlier authorities. Nevertheless, when he finally reached his conclusion, his view, and not the view of the earlier authorities, became the law.[2]

This extraordinary description of the rule "the law is like the later" is far from a unanimous, or even a majority view. Though currently popular, it is based on the opinion of only some of the later halakhic authorities. While, from a legal point of view, only the rule's final formulation is valid (an interesting case of halakhic self-reference), it is nonetheless illuminating to inquire into its early history and the meanings attributed to it in the past. Although a thorough investigation of the subject would require specialized historical expertise, even a schematic outline should suffice to stimulate reflection on the true significance of "the fundamental principle of decision making in Jewish law" and help determine whether, indeed, it fits this description.

To the best of my knowledge, the earliest source that presents the rule in this spirit is R. Judah, the son of the Rosh, in a responsum (*Zikhron Yehuda* #23) in which he sought to rule according to the decision of his father against the author of *Sefer Haterumot* on the grounds that his father was a later authority. His opponent disagreed, arguing that "the principle of 'the law is like the later' refers only to debates among Amoraim." R. Judah argued as follows: "My father, of blessed memory, asserted that "the principle that 'the law is like the later' also applies to later generations." He refers to the Rosh's commentary on bSanhedrin 4:6, in which the Rosh employs the principle "Jephthah in his generation is like Samuel in his" and not the principle "the law is like the later authority." R. Judah's brother R. Jacob, author of the *Turim*, concurs, saying "in the commentary on Sanhedrin, my father, my master, of blessed memory, wrote that the law is like the later holds even among contemporary scholars" (*ibid.*). This development of the Rosh's teaching is the first time that the principle "the law is like the later" is referred to in the halakhic literature as bestowing legal primacy on decisions of an authority who is later than others who have ruled on the subject, as Elon has argued.

2 *Ibid.*, p. 271.

R.M. Alashkar quotes R. Joseph Colon of the fifteenth century:

> You have written that you were told by your teacher on the authority of the most honored R. Joseph Colon that the rule of "the law is like the later" applies to post-talmudic authorities (*poskim*) as well ... and that R. Joseph Colon went as far as to assert that the rule applies to the opinion of a single later authority when it opposes the opinion of many others who preceded him, for example, the author of *Sefer Hamitzvot Hakatan*.[3]

R.M. Alashkar attacks this view, ridiculing it:

> This view is groundless and does not even merit refutation. I wish I knew the period up to which he meant to say that "the law is like the later." If it was intended only up to a certain generation, then the rule is qualified and cannot be applied generally; furthermore, there is no criterion for preferring one generation to another. Nor can we imagine that he did not wish to qualify his statement and that he intended to extend the scope of the rule forever. This would be preposterous, for our later generations resemble those of the earlier Sages less than a man resembles an ape, and we could only wish for sense enough to understand their simplest teachings. All the more so for generations yet to come, since there is a gradual decline of intelligence.

He concludes:

> The Geonim who ruled that "the law is like the later" referred only to the later generations of the talmudic period that preceded them. They could hardly have been referring to future generations whose opinions they never heard and whose methods they were unfamiliar with, for only prophets can foresee the future.[4]

Indeed, from the time of R. Joseph Colon, Ashkenazic and Sephardic authorities differed explicitly and systematically over the scope and meaning

---

3 R.M. Alashkar, *Responsa*, #54. For more on R. Joseph Colon's view, cf. #84 and #94, discussed below. Israel Yuval interprets the remarks by the Rosh, which his sons used as grounds for their argument, somewhat differently: "The position attributed to the Rosh, i.e., that he used the principle even in deciding disputes between post-talmudic authorities, is a later interpretation of his stance, as is evidenced by an examination of his actual words" (Israel Yuval, "Rishonim Veahronim; Antiqui et Moderni," *Zion* 57 [1993]). This interpretation of the Rosh is groundless, for Yuval has failed to notice that what he calls "a later interpretation" is, in fact, the position taken by the Rosh's own sons, including R. Jacob, who studied with him for decades.

4 *Ibid.*, at the beginning of the responsum.

of the rule "the law is like the later." R. Moses Isserles of Cracow, the greatest of the later Ashkenazic authorities, clearly illustrates this when expressing his own position in the preface to *Darkei Moshe*. Explaining his motivation for writing, he gives as one of his reasons:

> Which is the essence and purpose ... because it is known that the learned author of *Beit Yosef* had a natural inclination toward the great ... Alfasi, Maimonides and Asheri ... even though they are early, not later authorities ... but he does not heed the proclamation of the earliest authorities ... instructing us to decide according to the later authorities and to disregard the earlier ones, even when a student rules against his teacher. And the later authorities, led by R. Joseph Colon and R. Israel Isserlein, always decided according to this ... so he opposes the customs of those countries which usually rely on this rule straightforwardly.[5]

R. Israel Isserlein,[6] R. Joseph Colon,[7] R. Jacob Pollak, R. Shalom Shakhna and R. Moses Isserles are, indeed, staunch supporters of the view that the rule refers to authorities of the post-talmudic period, including themselves, and extends "to the end of all generations." For them, the rule that "the law is like the later" had indeed become what Elon calls "the fundamental principle of decision making in Jewish law."[8] As we shall see, this strong interpretation originates in the fourteenth century. I can find no trace of it earlier, in either the Ashkenazic or the Sephardic authorities.

As to the Sephardic authorities, their tradition, represented by R. Joseph Caro,[9] R.M. Alashkar[10] and R.M. Alshekh,[11] clearly differed from that of the Ashkenazic *poskim*. They relied on the great early authorities, renowned for their wisdom, rather than on those of later periods. Maimonides states this doctrine in the introduction to his *Code*:

> Furthermore, if one of the Geonim taught that the law is so and so, and to another court of a later period it appears differently on the basis of the law as written in

5 Our translation. This passage is also cited in Elon, *op. cit.*, vol. 3, p. 1354.

6 Who ruled, for example, that R. Meir of Rothenburg is a "later authority" with regard to R. Isaac of Duran. See *Terumat Hadeshen*, part 1, #215.

7 See note 3.

8 See note 2.

9 In his introduction to the *Shulhan Arukh*; see Elon, *op.cit.*, vol. 3, pp. 1309ff., 1316ff.

10 See note 4.

11 R.M. Alshekh, *Responsa*, #39.

the Talmud – we are not bound to accept the earlier opinion, but the more convincing one, regardless whether it is that of the earlier or the later authority.[12]

The Ashkenazic authorities had a different policy. Elon[13] notes that R. Moses Isserles follows his teacher R. Shalom Shakhna, who refused his many students' request that he "become a *posek*," that is, commit his decisions to writing, saying:

> I know that they will not decide differently from what I would write, since the law is like the later, and I do not wish them to rely on me ... for the judge should only consider what he beholds with his own eyes. Therefore each one should judge for his own time at his own discretion.[14, 15]

---

12 Maimonides, *Code*, Rubinstein ed. (Jerusalem: 1956–7), p. 13. An additional Sephardic source in the same vein appears in R. Bezalel Ashkenazi's *Klalei Hatalmud*, as reprinted by A. Marx in the Hebrew section of A. Marx, *Festschrift zum Siebzigsten Geburtstage David Hoffmans* (Berlin: 1914), p. 209, note 306:

> From *Seder Olam*, written by Nachmanides, [Maimonides, notes Marx], it is known that the law is according to the later authorities to the extent that they reached a decision, whether they decided for the majority or the minority view, or between two opposing individual views. This does not mean that the later authorities dissented from the views held by the earlier authorities on any point whatever, God forbid.

The nature and content of Maimonides' *Seder Olam*, which is mainly known from quotes in R. Bezalel Ashkenazi's *Klalei Hatalmud*, are not clear. There is little doubt, however, that the work accurately represents the views of the Sephardic authorities on this matter.

13 Elon, *op. cit.*, vol. 3, pp. 1345ff.

14 R. Moses Isserles, *Responsa*, #25 (Ziv ed. p. 156b; our translation). This responsum by R. Israel, the son of R. Shalom Shakhna, is quoted in Elon, *op. cit.*, vol. 3, p. 1347.

15 This passage touches on a historically significant point which is pertinent to the issues underlying our investigation. The reason for the appeal by R. Shalom Shakhna's students that he become a *posek* is given at the beginning of the passage: "He trained many students from one end of the world to the other, and we all live by his teachings, which sustain us." This constituted yet another unsuccessful attempt to establish a central judicial authority for Ashkenazic Jewry. It failed for the same reasons the others failed, with one salient difference. For the history of these attempts, see E. Zimmer, *Jewish Synods in Germany during the Late Middle Ages* (New York: 1978).

As his son testifies later in the passage, R. Shalom Shakhna himself followed the policy of his own teacher, R. Jacob Pollak, on this matter: "and for the same reason his teacher the illustrious R. Jacob Pollak did not write any

---

The main reason for the sterility of these efforts was the tradition of independence in the early Ashkenazic judiciary, of the non-intervention of one community in the affairs of another. This principle is already in evidence at the time of R. Joseph Bonfils, who lived in France in the early eleventh century. See A. Grossman, "The Attitude of Early Scholars of Ashkenaz toward the Authority of the *Kahal*" (Hebrew), *Shenaton Hamishpat Haivri* 2 (1975), pp. 180–181. Throughout its history, Ashkenazic Jewry never had a "Chief Rabbi," an individual with ultimate judicial authority (*posek yahid*), or even a central authoritative rabbinical forum, except during the time of Rabbenu Tam.

On the other hand, the four major pillars of decision-making in Jewish law: Alfasi, Maimonides, Asheri and R. Joseph Caro, all flourished in lands where the communal social-political frameworks had always recognized the supreme authority of a central figure – a rabbinical authority who enjoyed royal recognition and official status. For a summary of the Sephardic legal literature, see H. Zimmels's *Codifications by the Jews of Spain (The Sephardi Heritage)* (New York: 1971). Zimmels noted that all four of these outstanding figures had some connection with Sephardic lands, but he did not attribute any significance to this fact; he also failed to notice that even though their connections with "Spain" were incidental, the fact that they all lived where there were centralized political frameworks has clear historical relevance. Such centralization was a necessary condition for the creation of authoritative codes of Jewish law.

Against the background of the highly decentralized community organization that had existed since ancient times in the Ashkenazic countries (as well as in society in general) it is almost impossible to imagine the creation of a central, authoritative code of Jewish law. All that was created there during the twelfth and thirteenth centuries was a series of summaries of the legal works of the great commentators, such as Rashi and the Tosafists, compiled and arranged by their students, along with the students' analyses of their teachers' halakhic opinions. See my "The Characteristic Features of Halakhic Literature in Ashkenaz in the Thirteenth and Fourteenth Centuries" (Hebrew), in *Alei Sefer* 4 (1977), esp. pp. 20, 32–34.

The difference between earlier attempts at organizing a central authority and that of R. Shalom Shakhna's students is that, conscious of the problems inherent in assenting to his students' request, R. Shalom Shakhna himself sought to preserve the principle of freedom of judicial decision-making for future generations. It should also be noted that implied – though not articulated – in this passage, is the beginning of a shift in the application of the rule "the law is like the later" from the last authority of the past (the traditional Ashkenazic interpretation) to the present authority. For more on this, see sec. 4 below.

books. Nor did they copy and put on file any of the responsa which they sent to distant places, because they considered this overbearing assertiveness."[16]

Elon's description is indeed an accurate one of this Ashkenazic understanding of "the law is like the later," which originated in the fourteenth century and has been dominant since. However, his account fails to acknowledge that the opinion of R. Joseph Caro and the Sephardic authorities, both before and after him, Maimonides included, is different. In their opinion, the rule does not apply to post-talmudic times at all.

Though I discuss the historical and theoretical origins of the dispute at some length, the divergence between the two traditions is not the main issue here. As we will see, there is no early evidence, and maybe even no late evidence, that the rule "the law is like the later" was meant to uphold contemporary decisions against a contradictory precedent. On the contrary, it tends to restrict rather than extend authority; it does not grant authority to the contemporary *posek*, but rather refers him to the authorities of the past with the stipulation that of the past authorities those of a later period must be preferred. That the current *posek* is also a "later authority" – in fact, the latest – is an entirely novel idea of Ashkenazic origin for which I can find no traditional sources, as I explain below. While the principle conferring authority upon the current *posek* is indeed genuine, it originates in the altogether different rule: "Jephthah in his generation is like Samuel in his generation." As we shall see, this has nothing whatever to do with the issue of "the law is like the later."[17]

2. The Rules "The Law Is Like the Later" and "The Law Is Not According to a Student in the Presence of his Teacher"

This question of its applicability to later generations of halakhic authorities only arose after the rule "the law is like the later" had been in existence for a long time, evolving considerably over the years. There is no trace of it in the talmudic literature, and it is first mentioned in *Seder Tanaim Veamoraim* of the Geonic period.[18]

16 R. Moses Isserles, *loc. cit.* (our translation); quoted in Elon, *op. cit.*, vol. 3, p. 1347.

17 Cf. section 3 below. Elon too considers the rule "Jephthah in his generation is like Samuel in his" together with "the law is like the later," but does not discuss the relationship between the two rules (*ibid.*, vol. 1, pp. 266).

18 The many print and manuscript versions, and numerous editions, of *Seder Tanaim Veamoraim*, is a serious problem requiring further study. The last critical

In the version published by R. Haim Joseph David Azoulay of *Shem Hagedolim*, the rule appears twice:

> Even though the Sages defined rules such as "In a controversy between A and B the law is according to A"; if authoritative opinions of later Amoraim exist, the law is as they decide.[19]

And he continues:

> In instances in which two Tannaim or Amoraim hold different views and it is not stated whose opinion is upheld, if one is a student who contradicts his teacher, then the opinion of the student is not preferred over the opinion of his teacher. This is the rule up to the time of Rava. From his time onward, the law is like the later.[20]

In the *Mahzor Vitry* version, the first rule is the same; the second ends with the words:

> The opinion of the student is not accepted against the opinion of the teacher.[21]

In the version published by Samuel David Luzzatto (Shadal), neither of R. Azoulay's rules is given. Rather, the first rule is as follows:

> Even though the Sages said "In a controversy between A and B the law is according to A," if there are existing authoritative opinions of our masters, the Amoraim, the law is according to them.[22]

---

edition is K. Cahane's (Frankfurt on Main: 1935). The classic work on the subject is still A. Marks's paper in *Tiferet Yisrael* (Breslau: 1911). For an updated comprehensive bibliography see Y. Efrati's *Tekufat Hasaboraim Vesifruta* (Petach Tikva: 1973), pp. 14–32). In Efrati's opinion, *Seder Tanaim Veamoraim* came into being as a unified work (the version in *Mahzor Vitry*) only when *Mahzor Vitry* was compiled from the various early manuscript versions then available. I have not tried to determine which of the existing versions is preferable, indeed, I doubt it can be done. See S. Abramson, "Letoldot Nusah Seder Tanaim Veamoraim," in *Ezra Tzion Melamed Jubilee Volume* (Hebrew), (Ramat Gan: 1982), pp. 215–248.

19 R.H.J.D. Azoulay, *Shem Hagedolim*, Maarekhet Sfarim, *samekh*, *Seder Tanaim Veamoraim*, sec. 24.

20 *Ibid.*, sec. 25.

21 *Mahzor Vitry*, p. 491.

22 S.D. Luzzatto, *Kerem Hemed* (1839), vol. 4, p. 193.

This is a different matter altogether (if not simply an error). The second rule is identical with the version in *Mahzor Vitry*. Nevertheless, there is no disagreement here, for Luzzatto, a few lines later, in referring to the Amoraim, himself says:

> In an instance where early Amoraim such as Rab and Samuel hold different opinions, and there are also opinions of later Amoraim such as Abbaye and Rava or others, the law is according to the latter,[23] and if earlier and later authorities voice different opinions [in R. Azoulay's version these words are added: "for instance, their students, who added to the words of the earlier authorities"[24]], the law is like the later.

Aside from stylistic and editorial variations, the different versions do not conflict with each other. This indicates that contrary to the common misconception, the statement "from Rava on, the law is like the later" is not a universally binding condition. It applies only to cases where teacher and student disagree, and tells us that in these cases it is permissible to decide against early Amoraim, including Rab and Samuel, on the basis of the rule "the law is like the later."[25]

Furthermore, a closer look at section 25 in Azoulay's *Seder Tanaim Veamoraim* reveals that originally, the condition "till Rava" did not apply to

---

23 Cf. R. Meir of Rothenburg (Prague edition: 1917) quoting *Halakhot Ketzubot*: They [the Geonim] have communicated from the Academy – when it is found that two Amoraim disagree on a matter and there are later Amoraim who debate the same subject, the law is like the later, and we have a tradition that there were six generations of Amoraim.
And see note 26 below.

24 R.H.J.D. Azoulay, *op. cit.*, sec. 51.

25 This obviates Ginzberg's argument (in his *Geonica*, vol. 2, pp. 21–22) attributing great significance to the Geonic responsum (*ibid.*, p. 32) which states that the law is according to Rabba b. R. Huna (of the third generation) because he is "later" than Rab, even though they both preceded the generation of Abbaye and Rava. There is nothing new in this, since Rabba b. R. Huna was not a student of Rab (emendation to "Rabba b. Hana" is useless because the latter was Rab's contemporary and cannot be described as "later"). *Tosafot* (Kidushin 45b s.v. *hava uvda*) quoted by Ginzberg is also irrelevant, since it deals with R. Jose – a student of Rab.
See Elon, *op. cit.*, vol. 1, p. 268, note 103, regarding the fact that Asheri (Baba Metzia 3:10; 4:19–21) expressed surprise at Alfasi's opinion on this matter. Elon argues that "Alfasi's statement, referred to by Asheri, indicates that this rule applies also to the period prior to Abbaye and Rava," implying that Asheri disagrees. This is incorrect. Asheri himself emphasized that "I could not comprehend the view of R. Alfasi ... because only from the time of Abbaye and Rava do we decide according to the later, but before that the law is not decided according

"the law is like the later," but to the other rule, "the law is not according to a student in the presence of his teacher." *Seder Tanaim Veamoraim* holds that this rule only applies "up to Rava"; from then on, "the law is like the later" – i.e., the student.[26]

Now the rule "the law is not according to a student in the presence of his teacher" is not found in the Talmud either; it originates in the Geonic period.[27] To be sure, there are numerous talmudic references to a strict prohibition against a student's ruling in the presence of – or within a certain geographical distance of – his teacher, as well as references to a different prohibition against giving an opinion before one is properly qualified. The Talmud regards this as a commonly-encountered problem with serious consequences.[28] It also deplores "dissent from the academy of one's teacher"[29] – an enigmatic expression – one of a number of behaviors showing disrespect toward teachers.[30]

None of this, however, implies that a qualified person is prohibited from ruling against the opinion of his teacher. On the contrary, some of the most renowned Sages took issue with their teachers when they found their teachers' views unacceptable. The deciding rule here is "if profaning of the holy

---

to a student against his teacher." Clearly, when a student–teacher controversy is not involved, Asheri agrees that the time of Abbaye and Rava does not serve as a boundary. Note that Alfasi rejected the time of Abbaye and Rava as a turning point, as his ruling at Baba Batra 34b, pericope *hahu arba*, shows.

I note these points only briefly, as my main concern is not with Asheri and *Tosafot* but with Geonic opinions. The fact is that in *Tosafot*, the rule is applied before the time of Abbaye and Rava even in a case involving a student–teacher dispute. See *Tosafot*, Rosh Hashana 34b s.v. *ledidi*: "Nevertheless since R. Abbahu disagrees [with R. Johanan his teacher!] we are inclined to rule according to him because he is later." Moreover, none of the passages quoted in the *Encyclopedia Hatalmudit* s.v. *hilkhata kebatrai* (the law is like the later), in support of the view that the rule was applied before Abbaye and Rava, involve a student–teacher controversy. I have no knowledge of anyone disagreeing with this.

26 Apparently S. Assaf holds the same view; see his "Darkei Hatalmud Uklalei Hahoraa Beteshuvot Hageonim" in *Tekufat Hageonim Vesifruta*, ed. M. Margalioth (Jerusalem: 1955), secs. 41–42.

27 R. Hananel, Sanhedrin 22a.

28 See, e.g., bSanhedrin 5b.

29 bBerakhot 27b.

30 For details, see M. Aberbach, "The Relations between Master and Disciple in the Talmudic Age," in *Exploring the Talmud*, ed. H.Z. Dimitrovsky (New York: 1976).

name is threatened, the respect due a teacher need not be accorded."[31] There is certainly no sign of a "decision rule" in the Talmud to the effect that "the law is not according to the student in the presence of his teacher." However, the Talmud does state that although "the law is according to a third authority who tips the scales," this rule does not apply to an authority deciding against his teacher.[32] Indeed, R. Hananel offers this interpretation of the rule, "the opinion of a third authority is not decisive,"[33] to make it compatible with the previous rule, but we cannot infer from this that the law is always decided in accord with the teacher's view rather than that of the student. The rule cited by R. Hananel has to do with whether, in a dispute between a teacher and another authority, the views of a student who is somewhere in between can be used to break the stalemate and create a majority. The opinion of the student is regarded as "dependent" on the opinion of his teacher, and cannot be counted toward a majority vote on a point on which they agree.[34] In the particular case in question, it is the opinion of R. Akiba that does not tip the scales in favor of either the opinion of R. Eliezer or that of R. Joshua, and R. Akiba is a student of both.[35] Clearly, in a dispute between two unrelated individuals, the law is not necessarily decided according to the teacher.

Be that as it may, the rule "the law is not according to the student in the presence of his teacher" was formulated by *Seder Tanaim Veamoraim*, which is therefore at liberty to qualify its account. It does just that by adding the limitation "till Rava," specifying that after Rava "the law is like the later" – even when it involves a student's ruling against his teacher. But when a student–teacher relationship is not involved, the law is like the later even before the generation of Abbaye and Rava. This is clear from perusal of the other places in *Seder Tanaim Veamoraim* where the rule is used: it is stated generally, without specifying a *terminus a quo*.[36] This observation appears to be

31 bBerakhot 19b.

32 bShabbat 39b.

33 bPesahim 21a.

34 R. Eliezer b. Nathan (section 6) offers a novel interpretation of these passages and R. Hananel's remarks. It suggests that the student's opinion is particularly weighty because he is familiar with his teacher's opinions and their rationales. But this interpretation seems to conflict with the plain meaning of the texts.

35 On the possible existence of a version which reads "We do not learn the law from students," see S. Abramson, "Birurim," *Sinai* 58 (1966), pp. 183–185.

36 This is not unusual in the halakhic literature. The author of *Yad Malakhi*, for example, wrote "I saw the new interpretation by Damesek Eliezer, which in his

verified by a statement of R. Hai Gaon cited in one of R. Joseph Colon's responsa:

> And R. Hai Gaon wrote as follows: and now we examine the statement of R. Johanan who is a later authority; do we therefore assume that he heard the statement of Rabbi [Judah the Prince]: "authorize expenses for maintenance" (bKetubot 69a), found it unacceptable, and rejected it, and since he is a later authority, we do not follow the opinion of Rabbi, or perhaps … .[37]

R. Hai Gaon applies the rule "the law is like the later" quite casually, then, to one of the earliest Amoraim, and even uses it against R. Judah the Prince; neither he nor R. Joseph Colon have any problem with this. I should add here that though R. Johanan relates that he attended the lectures of R. Judah the Prince,[38] this does not make R. Johanan his student, because he was too young at the time, and – as he admits – could not follow the discussions at the academy.

For reasons I explain in section 4 below, there are serious doubts about whether R. Hai Gaon indeed made the statement attributed to him by R. Joseph Colon, and no additional source quotes or refers to the responsum in question. There is no doubt, however, that in substance, it reflects the opinion of R. Hai Gaon. The relevant material on this point has been collected by Groner,[39] who reports that there is no known instance in which R. Hai Gaon invoked the "student against his teacher" rule in deciding a controversy among later Amoraim. "The last authority to whom R. Hai Gaon applies this rule is R. Hisda." Groner also asserts: "but … for those earlier generations R. Hai Gaon did not invoke the rule 'the law is like the later.' He applies the rule only from Rabba and R. Joseph on."[40] This is incorrect, even if we deny the authenticity of the responsum quoted by R. Joseph Colon – which Groner does not do. Here Groner is influenced by R. Yomtov of Seville (Ritba), who writes:

> The accepted rule that "the law is not according to a student in the presence of his teacher," only applies if they differ during the lifetime of the teacher, that is,

---

opinion also applies to the time of the Tannaim." Cf. R.A. Efrati's comment in his *Even Tzion* (Warsaw: 1866), which includes "a comprehensive study of *Seder Tanaim Veamoraim*" (p. 31).

37 R. Joseph Colon, *Responsa*, #94. This quote from R. Hai was apparently unknown to B.M. Lewin, who does not include it in *Otzar Hageonim*.

38 bHulin 137b.

39 Z. Groner, "Rav Hai Gaon Vedarko Bahalakha" (PhD diss., The Hebrew University, 1974).

40 *Ibid.*, pp. 67–68.

> if the teacher heard the dissenting view and did not change his opinion. However, if the student disagreed with his teacher after the latter's death this rule does not apply, otherwise we could not uphold the "like the later" rule, since every later authority is a student of an earlier authority – and the law is not according to the student against the teacher! And this is confirmed by R. Hai Gaon, may he rest in peace.[41]

On this analysis, all "later authorities" are in principle "the students of earlier authorities" and the distinction we made earlier is lost. But the Ritba is the only source attributing this view to R. Hai Gaon, and it is impossible that he held it, because he clearly states elsewhere that when the law is decided according to a teacher against a student it also applies to the students of the students,[42] who surely arrived at their opinions after the death of the first teacher, and not in his presence.

I am thus certain that Groner's conclusion – "Apparently, 'the law is like the later' [in his opinion, in later generations only] is applicable only when the student did not differ from his teacher in his lifetime" – is untenable. The doctrine of R. Hai Gaon is quite the reverse, since in almost every case he would state that the law is like the later, as explained above.

The Ritba's doctrine[43] is interesting in its own right, however, and constitutes proof that R. Hai Gaon does not hold the same opinion. It should be noted that, as he himself states, the Ritba's main concern is the apparent conflict between the two rules "the law is like the later" and "the law is not according to a student in the presence of his teacher," since a student is by definition "later" than his teacher. As we have already seen, the two rules do not contradict each other. On the contrary, *Seder Tanaim Veamoraim*, the main source of these rules, explicitly links them, stating that one applies up to the time of Abbaye and Rava, and the other thereafter. This seems to indicate that the Ritba based his formulation and interpretation of "the law is like the later" on a source other than *Seder Tanaim Veamoraim*.

Further scrutiny reveals that the Ritba's view is shared by the Rashba,[44] Ran[45] and *Nimukei Yosef*, quoting R. Aharon Halevi.[46] These views are

---

41 *Otzar Hageonim*, Nedarim 75a.

42 See his remarks in Harkaby's *Teshuvot Hageonim*, sec. 358 (700, vol. 2, p. 43).

43 See the text near note 41.

44 *Hidushei Harashba*, Shabbat 141b.

45 Rabbenu Nissim's commentary on Alfasi, the beginning of Suka.

46 R. Joseph Habiba's commentary on Alfasi *ad loc.*

summarized by R. Bezalel Ashkenazi in *Klalei Hatalmud*.[47] This is also the view of Nachmanides, namely, that the law is not according to a student against his teacher even if the student is a "later" authority.[48] It is clear to me that these four fathers of the halakha did not preserve the "like the later" rule in its original form as stated in *Seder Tanaim Veamoraim*,[49] but rather, adopted Alfasi's interpretation of it. In Alfasi's opinion "the law is like the later" applies even before the time of Abbaye and Rava and even if a student disagrees with his teacher. In following Alfasi's interpretation, these authorities were compelled to resolve the apparent paradox which arises when the two rules – "the law is like the later" and "the law is not according to the student in the presence of his teacher" – are taken together.

It can be stated with certainty that this problematic interpretation was not adopted by R. Hai Gaon. In any case, whether one accepts Groner's conclusion that R. Hai Gaon does not apply the rule before Abbaye and Rava, or my view that he applies the rule partially (that is, not in teacher/student conflicts) before Abbaye and Rava, there is no contradiction between the two rules, which are fully compatible and even complementary. Hence there is no need for the artificial solution attributed to R. Hai Gaon by the Ritba, or for any other such contrived compromise.

3. "Jephthah in his Generation Is Like Samuel in His"

As is the case with all halakhic rules, the scope of "the law is like the later" is limited by its context. It stipulates that when a *posek* has to decide between conflicting opinions of earlier and later authorities, and there are no other relevant considerations or decision-making rules, such as majority–minority rulings, religious prohibitions, or monetary law, he is not merely permitted, but obliged, to prefer the later opinion. Thus, "the law is like the later" is in no way a principle delegating decision-making power or jurisdiction to the current *posek*. On the contrary, like all halakhic guidelines, the binding directive "like the later" deprives the *posek* of his freedom of independent decision-making. It certainly does not establish him as the "last of the later."

47 R. Bezalel Ashkenazi, *op. cit.* (see note 12), no. 311, p. 204.
48 The opinion of Nachmanides is quoted by R. Solomon b. Adret (Rashba); see note 44.
49 See note 25.

At the same time, there is no doubt that even as early as the talmudic era, the responsibility for making final decisions had been delegated to the contemporary legal authority or court. Not by virtue of the rule "the law is like the later," which was unknown at the time, but on the basis of a more fundamental principle: "Jephthah in his generation is like Samuel in his."[50]

In its original context, this principle is concerned not with resolving controversies, but with the delegation of power and authority to the courts as binding judicial institutions. The rationale for it is found in the Mishnah:

> And on another occasion two came and said, "We saw it at its proper time but on the following night it could not be seen," and Rabban Gamaliel accepted them. R. Dosa b. Hyrcanos said, "They are false witnesses, how can people testify that a woman has delivered if her belly is between her teeth on the following day?" R. Joshua said: "I agree with you." Rabban Gamaliel sent a message to him: "I command you to come to me with your walking stick and money on the day which according to your calculation is the Day of Atonement." R. Akiba went to him and saw that he was troubled... . He went to R. Dosa b. Hyrcanos who said to him, "If we were to argue against the court of Rabban Gamaliel, we would have to argue with the rulings of every court which has existed from the time of Moses... but any three persons who have arisen as a court over Israel are like the court of Moses."[51]

The Talmud explains: "even the least worthy person, if appointed to public office, is like the noblest of the noble." It is interesting to note the descending order here: "the court of Moses," "any three persons who have arisen as a court," "the least worthy person." As it indicates, the Sages regarded all courts as equal with respect to judicial authority, in spite of their belief that mankind had degenerated from the days of Moses onward, a process reflected in the descending order of the courts noted by the Talmud. This dilemma did not go unnoticed by the Sages, who stated their position concisely at the end of the pericope:

> "And you shall come to the priests and Levites and judges who shall be in those days."[52] Could someone go to a judge who did not live in his own time? Of course not. That is, "Do not say those of the past were better than these."[53]

---

50 bRosh Hashana 25b.

51 Mishnah, *ibid.*, 2:8–9.

52 Deut. 17:9.

53 Eccl. 7:10.

The Talmud thus explicitly rejects evaluation of contemporary courts by comparing them with courts of other eras, though it does not explain why such speculation is prohibited.

There is no direct connection between the delegation of power to courts and the rule "the law is like the later." As we saw, the latter has to do with the procedure for deciding points on which earlier *poskim* did not agree, and obligates the contemporary *posek* (note, however, that courts also fall into this category) to adopt the more recent ruling. The contemporary court cannot, however, include itself among the "later," and is not free to follow one of the earlier authorities, even if it prefers that opinion, since "earlier" and "later" refer only to authorities of the talmudic era. Thus the thrust of "like the later" is that contemporary courts do not enjoy judicial autonomy as far as talmudic disputes are concerned.

The delegation of judicial power, on the other hand, places no restrictions or obligations on the court, but only on those under its jurisdiction. By its very nature it concerns the present – contemporary courts adjudicating current disputes – enabling courts to exercise authority when they deviate from halakhic precedent.

Delegation of authority to contemporary courts can be viewed, in a trivial sense, as covering all current disputes, including those over talmudic precedent where the "like the later" rule would be a relevant decision-making guideline. However, this does not suffice for the Geonic rule "the law is like the later" to be automatically derived from it. From the principle of the delegation of judicial power it does not follow that future generations must also abide by the decisions of the court of the present generation, since they will have their own courts; hence, even if contemporary courts favor later authorities, the delegation principle alone cannot ensure that successive generations do not ignore them.

The principle "the judge who shall be in those days," and the rule "the law is like the later" are neither identical nor parallel, but at the same time, they are not contradictory. Their relationship may be described as complementary; they complement each other by limiting each other.[54]

---

54 The problems that arise from the delegation of jurisdiction to the courts are illustrated by the apparent contradiction between this principle and Mishnah Eduyot 1:5, which states that "one court cannot annul the ruling of another until it excels it in wisdom and numbers." Maimonides resolves this contradiction as follows:

> If a Great Court, using the proper procedure, decided on a law and another court after it has reason to contradict it, then the second court adjudicates as

The Amoraim recognized independent jurisdiction as an animating, regenerative principle in Jewish law. The decisions of the past, which they regarded as an era superior to their own, did not have the power to bind current judicial institutions, (though in practice they usually followed precedent), and their own rulings lacked the power to constrain future halakhic developments. A similar principle was applied to rulings of individual authorities, outside the courts, with the reservation that students were required to pay their teachers proper respect. Even within the area of individual rulings, autonomy in decision-making was recognized, as is reflected in the rule, "when the holy name is threatened, the respect due a teacher need not be accorded."

The Geonim added to the principles of their predecessors, asserting that by virtue of the power to resolve conflicts and even annul previous rulings that had been conferred on successive generations of "present" authorities, "the present" was endowed with permanent jurisdiction on questions relating to the past, though not the future.

The new element here is thus the retrospective or diachronic dimension added to the courts' prescribed authority. In principle, the Geonim declared, courts could decide according to their needs; they could exercise discretion.

> it sees fit, since it is written "to the judge who will be in those days." One is bound only by the court of his own generation.

But,

> A court that issued a decree or regulation or established a custom ... and a later court wished to revoke it ... cannot do so until it exceeds the first court in wisdom and in numbers (*Code*, Laws Concerning Rebels 2:1).

In his commentary, *Kesef Mishne*, *ad loc.*, R. Joseph Caro notes that as a matter of historical fact, a categorical distinction was indeed invariably made between the Tannaitic and Amoraic periods, a situation which seems problematic given Maimonides' solution. Hard pressed, he explains:

> We can say that from the time of the redaction of the Mishnah, later generations undertook not to contradict their predecessors, and the same thing occurred after the redaction of the Talmud – from the time of its redaction, no one has been permitted to differ with it.

For rabbinical material on the subject (including "the law is like the later") see A.Z. Ben Zimra, "The Court Cannot ... " (Hebrew), in *Kol Sinai* (1964).

The Geonim certainly did not dispute the talmudic authority-delegating rule "Jephthah in his generation is like Samuel in his," but added that a court seeking to decide a controversy between early authorities in a halakhically correct manner should always prefer the later authority among the contending views, because in that authority's own time the law was decided thus. This relative advantage of the later of the contending past decisions remains for the future, lending an element of "eternity" to each stage in talmudic decision-making. At the same time, though, the delegation principle ensures that future courts are free to make their own decisions.

The rule "the law is like the later," in this, its original form, is a purely formal-technical rule applying only to talmudic controversies. Its thrust is that every court, at the point in time when it is in session, has absolute and final authority to decide against all preceding courts – though in practice, as we said, courts would rule in most cases according to precedent – and this relative authority remains throughout future generations. There is nothing to suggest that, having formulated it, the Geonim applied it to contemporary courts. There is no question of a qualitative advantage of later rulings over earlier ones (such as their being "greater in wisdom"), thus querying whether the last Amora was aware of all the precedents misses the point entirely and is irrelevant to the issue as described here. All queries of this kind date from the post-talmudic period, after the rule had been "extended" to apply to this period as well. They arose as a result of the profound change that took place during this period in the interpretation and application of "the law is like the later".

4. The Turning Point in the Middle Ages

The Middle Ages saw a turning point in the history of "the law is like the later." Essentially, it had to do with the growing awareness on the part of the educated circles of twelfth century Europe that despite the unquestioned superiority of the ancients in every sphere: wisdom, morals, knowledge of the esoteric, proximity to God, and so on, contemporary intellectual achievements also deserved recognition. They rejected the earlier assumption that everything had been known to the ancients and that anything ignored by the ancients could be dismissed as unworthy of consideration. The rationale offered by medieval thinkers for this revolutionary premise was the aphorism of the dwarf on the shoulders of a giant: granted, the ancients were all intellectual giants, with infinitely wide horizons, and scholars of the twelfth century were mere dwarfs in comparison; nevertheless, by assiduous learning, these scholars could access the heritage of the ancients and sit, as it were, on the shoulders of giants.

This aphorism, with its deep philosophical, religious and cultural significance[55] was conceived by the philosopher Bernard of Chartres (d. 1126), one of the foremost Scholastics and head of the celebrated school in his city. Preserved in the works of his student John of Salisbury,[56] (d. 1180), it is typical of the peculiar genius, intellectual fertility, and intrinsic modesty that characterized the renaissance of the twelfth century.[57]

In Jewish writings the aphorism first appears in a responsum by R. Isaiah Di Trani, author of *Tosafot Rid*, who died in 1250.[58]

> Regarding what you have written that I must not contradict the great authority of R. Isaac, may he rest in peace. Heaven forbid that I should do such a thing. ... I am nothing but a flea ... yet I say that a statement which I cannot accept, I do not heed, even if it was uttered by Joshua the son of Nun, and I do not refrain from speaking my mind about it as far as my limited wisdom permits. ... and my witness is in heaven, that even when I think that I argue rightly against one of our early teachers, heaven forbid that my heart is vain to say that it was my wisdom that supported me, but I accept the metaphor which I learned from the wisdom of the philosophers. They asked the greatest philosopher among them: It is a given that the ancients were wiser and more knowledgeable than we are, but also, that we often dispute and contradict them, and we are right. How can this be? He replied as follows: Who can see farther, a giant or a dwarf? Surely the giant, because his eyes are higher than those of the dwarf. But if the giant carries the dwarf on his shoulders – who can see farther? Surely the dwarf, whose eyes are now above the eyes of the giant. We too are dwarfs riding on the shoulders of giants, because we have seen their wisdom and we transcend it. And it is by virtue of the power of their wisdom that we have learned all that we say, and not because we are greater than they were. My point is that we should not discuss the teachings of our early Sages except when we see that they disagree with each other: one forbids what another permits – whom shall we rely on? We cannot measure them on a scale ... and declare that since one is greater than the other his opinions should cancel out those of the other.

55 See G. Sarton, "Note," in *Isis* 24 (1935), pp. 107–109.

56 In his *Metalogicon*, ed. C.C.J. Webb, vol. 3, sec. 4, p. 136. The history and significance of the aphorism are rigorously treated in R.K. Merton, *On the Shoulders of Giants* (New York: 1965).

57 The classic work on the subject is C. Haskins, *The Renaissance of the Twelfth Century* (Cambridge, Mass.: 1927).

58 R. Isaiah Di Trani, *Responsa*, #62.

R. Isaiah Di Trani's acknowledgement – "I learned from the wisdom of the philosophers" – is interesting indeed. In fact, his source is none other than John of Salisbury's treatise, written about 50 years earlier. The first[59] to use R. Isaiah Di Trani's version of the aphorism was one of his students, R. Zedekiah b. R. Abraham, in the introduction to his *Shibolei Haleket*.[60]

The revolutionary dimension of the new intellectual climate symbolized so clearly by this aphorism is its recognition of the cumulative nature of learning, of the fact that a scholar's knowledge is not a self-contained entity, but a link in a chain. The more links added to it, the higher its quality. This awareness allowed medieval scholars to admit that the venerated ancients had sometimes been wrong – without detracting from their greatness and status as the true sources of wisdom or necessitating revision of the credo, "from their wisdom is ours".

Before this watershed, revisions and corrections could be justified only on the basis of the principle "their ancestors left them room to draw their own boundaries,"[61] which conflicts with that of "whoso breaketh through a fence, a serpent shall bite him" (Eccl. 10:8), which is said of those who deviate from earlier rulings (bAvoda Zara 27b). This principle has nothing to do with the nature of knowledge; rather, it recognizes a charitable attitude on the part of the ancients (or divine providence) toward future generations, lest their creativity be blocked. As Rashi says, "If our descendants find nothing on which to improve, how can they achieve fame?"[62] The Sages used this principle to assert: "a scholar who formulated a law [that is innovative – Rashi[63]] is not ignored."[64] Without this extraordinary principle, which acknowledges the theoretical possibility of successive generations having something to add to the teachings of their forebears, the authorities would have had to suppress any new idea.[65]

59 The Jewish versions of the aphorism are treated in D. Zlotnik's "The Commentary of R. Abraham Azoulay on the Mishnah" (Hebrew), *Paajr* 40 (1972), pp. 147–168 and "The Origins and History of the Aphorism of the Dwarf and the Giant" (Hebrew), *Sinai* 77 (1975), pp. 185–189.

60 *Shibolei Haleket Hashalem* (Vilna: 1887; Jerusalem: 1962), p. 18.

61 bHulin 7a.

62 Rashi *ad loc.* s.v. *makom hinihu* .

63 Rashi *ad loc.* s.v. *mikan*. Cf. *Shita Mekubetzet* (Vilna ed., sec. 1).

64 bHulin 7a.

65 My presentation here is intentionally simplified to highlight the basic difference between the received view prior to the twelfth–thirteenth centuries and that which gained ground subsequently. In reality, the Sages' attitude toward their

The eleventh and twelfth century French-Ashkenazic commentators on the Bible provide a good illustration of this approach. In commenting on a verse in Ecclesiastes – "Say not 'How was it the former days were better than these?' for it is not out of wisdom that thou inquirest concerning this"[66] – we find the following exchange. Rashi explains:

> Because the first generations were better and more pious than those of today, therefore former days were better than these because it is impossible that these days should be like former days.[67]

The Rashbam comments:

> Do not ask, what deeds are done in the world now ... that former days were better and more refined in their actions, because you do not ask this out of your inherent wisdom but because of your great stupidity because you should understand

---

predecessors is much more complex. It is marked by a differential evaluation of the various aspects of their accomplishments, though a parallel evaluative scale is used. Thus, for example, it did not go unnoticed by the great Amoraim that the quality of Talmud scholarship had been raised, not lowered, by their work, but they did not see this improvement as having any bearing on questions of authority, or any moral significance. Consider the interesting conversation between R. Papa and Abbaye:

> R. Papa said to Abbaye, How is it that miracles were performed for the ancients but not for us? It is not due to their learning. During all the years of R. Judah they only studied tort law, but we study all six branches of the law. And if R. Judah came to a passage about the laws of purity, he would say, This is a matter requiring deliberation for [our predecessors] Rab and Samuel, while we have twelve lectures on this subject at the academy. Nevertheless, when R. Judah took off one of his shoes [in preparation for a fast] the rains came, while we torment ourselves and cry out to heaven and no one listens to us. The other replied: The ancients suffered martyrdom willingly to sanctify the holy name ... (bBerakhot 20a).

Cf. the version in bTaanit 24b, which attributes this to the misbehavior of the entire community rather than to the shortcomings of its leaders. To adequately understand the policy of "orthodoxy" adopted through the ages, thorough familiarity with these issues is required.

66 Eccl. 7:10.

67 Rashi, Rosh Hashana 25b s.v. *al tomar*.

> by yourself from the evidence of your own eyes that the world deteriorates everyday.[68]

Another view of this deterioration is expressed by the twelfth-century scholar, R. Joseph Kara, in his commentary on the same verse:

> Do not ask what was in ancient days which were before us. If one tells you so-and-so learned the Torah so well that there is no other like him since then, do not say "because there is no wise man in this generation like him, I do not go to learn the Torah from a scholar of these days." And so it is with regard to personal conduct. Upon being told that in ancient times so-and-so sowed and reaped a hundredfold, or a man reared a young cow and two sheep and of the abundance of milk they gave he ate butter,[69] or a man exerted himself little and earned much, do not ask "why were previous generations so rewarded?" If you hear words like these do not say "Since the world is different I shall not toil any more." Do not say so, but as you conceive the world you should exert yourself in learning the Torah and toil for your sustenance.

A good example of this attitude is found in remarks in *Tosafot* on the passage in the Gemara which states that the judges of Caesarea erred in geometrical calculations pertaining to the ratio between a square and the circle that circumscribes it:

> It is perplexing – how could such men err ... given that they did not measure, how did they formulate a rule without any data to base it on? And we have to state that the rule "a square enscribed in a circle – half," was masoretic; the rule is indeed accurate, but refers to area rather than circumference (bSuka 8b s.v. *ribua*).

That is, they relied on a tradition and hence did not calculate it themselves. The tradition itself was accurate, but the judges of Caesarea erred in its interpretation.

In contrast to this traditional understanding, the new doctrine of knowledge as cumulative recognizes the possibility of ascent to greater heights, to a

---

68 Even those who deny the attribution of the commentary on Ecclesiastes to the Rashbam agree that it "includes many of the Rashbam's explanatory remarks on Ecclesiastes, and the debate should focus ... on one point only – whether the work in its present form was written solely by the Rashbam or also includes the commentaries of another scholar, who emended some of the Rashbam's comments." See A. Grossman, "The Commentary on Ecclesiastes Attributed to the Rashbam" (Hebrew), *Tarbiz* 45 (1976), p. 336.

69 Isaiah 7:21–22.

vantage point from which horizons are broader, allowing access to information that will, in turn, add to the stature of future generations. This insight marks a great step forward in the history of learning, though it does not even hint at the possibility of challenging the authority and wisdom of the ancients, let alone call for revolt. The crucial step in this direction would be taken only in the seventeenth century, with the Enlightenment's demand for a break from the binding heritage of traditional learning as preserved and handed down from ancient times. This would make possible discovery of the true order of the world through personal experience and observation of nature itself.[70]

Underlying the aphorism of the dwarf and the giant is the assumption, for the first time, that decisions of the "later authorities" are qualitatively superior to those of their predecessors. This made possible application of the rule to the post-talmudic period, transforming it from a decision-making tool useful only for adjudicating talmudic disputes into a fundamental legal principle. This breakthrough was achieved without detracting from the lofty status of the earlier authorities which to this day remains a cornerstone of the halakhic literature.

The novel epistemology expressed in the aphorism of the dwarf and the giant enabled the next historical development of the rule: inclusion of the *posek* and his contemporaries in the category of "the later." This could not have taken place without the elapse of a minimal period of time for the dissemination of the ideas underlying the aphorism and the internalization of their significance.[71] Thus, it could not have taken place before the fourteenth century.

Indeed, it took longer, and before the fifteenth century we have no example of the application of the rule "the law is like the later" to the post-talmudic period. The time-gap is recognizable from the noteworthy fact that there is no evidence the rule was ever applied to the long Geonic period. This would have required an established chronology of the Geonim and their succession

---

70 See R.A. Nisbet, *Tradition and Revolt* (New York: 1968).

71 R. Isaiah Di Trani did not embrace the new doctrine overnight. In his commentary on bBeitza, he explains the statement, "Now we are skilled in astronomical calculations," as follows:

> This does not mean that we are wiser than the ancients because we can calculate the time of the new moon and they could not. The ancients, though far more skilled in calculation than we are, were not permitted to act except on the basis of seeing the new moon.

as a prerequisite for accepting a particular opinion. As a rule, the Geonim themselves did not identify their predecessors by name in their decisions, though they did so in exceptional cases. Nor was there a binding chronology of the post-Geonic rabbinical authorities. The first to pursue questions of chronology, as far as I know, was R. Meir Hacohen, the fourteenth-century author of *Hagahot Maimoniot*, who emphasizes several times that R. Meir of Rothenburg is a "later authority" (relative to the Tosafists). It is, however, possible that the rule was applied with respect to the post-talmudic authorities earlier. See the cautious comments in a responsum by R. Abraham b. R. Isaac, the head of the Narbonne rabbinical court:

> And we rely on all he [the Rif] says, for he knew all that was said by preceding authorities, and differentiated the important points from the less important, as we say, the law is in accordance with the opinion of R. Ashi, because he is later (Kapah ed. [Jerusalem: 1962], #35).

More importantly, one must remember that the aphorism of the dwarf and the giant was a product of Christian scholasticism and part of its general worldview. To the best of my knowledge, the aphorism and the ideological compromise it entails were unknown in the fourteenth and fifteenth centuries among Islamic scholars, who also venerated "the glorious past."[72] Thus, they could not have influenced Jewish scholars in Islamic countries. This left its mark on Sephardic judicial decision-making up to the time of R. Joseph Caro, as we saw.

The new conception of the rule "the law is like the later" brought with it for the first time a demand that the decisions of the "earlier authorities" be made accessible to the "later authorities," since only in this way could the later authorities be visualized as riding on the shoulders of their predecessors. This demand – a basic element of "the law is like the later" for the late Ashkenazic authorities – is not mentioned at all in early sources, and first appears in the responsa of R. Joseph Colon, who cites a responsum of R. Hai Gaon.[73] As I noted above,[74] even if R. Hai Gaon indeed held the opinion attributed to him, the authenticity of the specific text, as quoted, is doubtful.

72 See S. Pines, "La Philosophy dans l'economie du genre humain selon Averroes: Une Reponse a Al-Farabi," in *Multiple Averroes* (Paris: 1978). I am grateful to M. Idel for this reference.

73 See above, note 37ff.

74 Note 39ff.

The problem relates to the question of when chronology became an issue. The requirement that "later authorities" be familiar with the decisions of their predecessors is not mentioned by any authority before R. Joseph Colon, and, indeed, given their method, was unnecessary.

Once this requirement was adopted, however, it served as the basis for R. Joseph Colon's stipulation that "the law is like the later" only applied

> where the opinions of the earlier authorities are written down in a well-known book [it can and indeed must then be presumed that they were known to the later *posek*]; however, what is written in a responsum of a Gaon but not mentioned by a well-known book, even if there is a later authority who decided against the opinion of the Gaon, it can be argued that perhaps this later authority did not know this and had he known it he would have retracted his opinion.[75]

R. Moses Isserles concurs.[76]

5. Conclusion

The perception of the rule "the law is like the later" as a fundamental principle of Jewish law reflects neither its original substance nor the spirit in which it was understood through most of its history, but only the last stage in its evolution according to the Ashkenazic tradition. The rule is basically just one of a number of guidelines, the majority of which were formulated in the Geonic (or Saboraic) period as part of the effort to regulate the normative use of the Talmud as an exclusive legal code. Like the other rules, it was meant to guide the *posek* or court in settling talmudic disputes. It most certainly should not be regarded as a law granting authority to the *posek*, or allowing his own inclusion in the category of "the later."

The principle of judicial independence, as applied to the rules for ongoing and exigency decision-making, was always a central element in halakhic practice, but it has never had anything to do with the rule "the law is like the later," which by definition limits rather than extends authority. The theoretical significance of the rule "the law is like the later," if it has any such significance, is to preserve for the future the relative authority of every stage of Amoraic ruling vis-a-vis rulings which preceded it. The Ashkenazic authorities of the

75 R. Joseph Colon, *Responsa*, #94 (see above, note 37).

76 R. Moses Isserles on *Shulhan Arukh*, Hoshen Mishpat 25b; also cited by Elon, *op. cit.* (see note 1), vol. 1, p. 271.

fourteenth apparently began to apply the "like the later" rule to the post-talmudic period when they started attributing to it a meaning more profound than that associated with other decision-making rules. Characteristically, such rules apply only to the Talmud, and are limiting in nature. An example is the rule "the law is according to the majority."

From the fifteenth and sixteenth centuries on, the Ashkenazic authorities began to apply "the law is like the later" to rulings of contemporary courts (though they did not admit this explicitly). This transformation came about as a result of changes that had taken place in the intellectual world of Christian Europe during the twelfth century (which Jewish scholars became aware of in the following century) with regard to the relationship between the wisdom of the ancients and contemporary scholarship. Its essence lay in the growing acceptance of the conception of wisdom as cumulatively acquired and developing from generation to generation. Poignantly encapsulated in the aphorism of the dwarf and the giant, this change was not internalized by the Islamic world, which accounts for the long period that was to elapse before the new interpretation of "the law is like the later" was accepted by Sephardic Jewish scholars.

6. Postscript (1994)

On the basis of the arguments presented in this paper, Israel Yuval has recently attempted, in his "Rishonim Veaharonim: Antiqui et Moderni," *Zion* 57 (1993), to arrive at a significant historical conclusion. The question he undertook to answer was as follows: At what point did the Ashkenazic scholars begin to recognize a boundary marking the close of the era of the Early Authorities and the beginning of the modern period? On the basis of my article, Yuval concluded that for the Ashkenazic community, the dividing line must be drawn during the Black Death and the disasters which came in its wake, that is, in the mid-fourteenth century. He contends that though the Maharil was the first to actually employ "the law is like the later," followed by the Mahari Weill and R. Israel Isserlein, its dissemination can be attributed to its use by R. Joseph Colon. As noted above, Yuval's interpretation of R. Asher b. Jehiel's sons' comments on the subject is different from my own; see note 3 above.

But according to Yuval, even R. Joseph Colon only extended the principle to authorities who lived and worked in the period between R. Meir of Rothenburg and R. Jacob, author of the *Turim*, that is, until the mid-fourteenth century. It is only this circumscribed group of scholars that the Maharil and his colleagues, followed by R. Jacob Colon, described as "later," relative to the authorities who preceded them. They did not apply the term to scholars who lived in the period between the mid-fourteenth century and their own period.

Yuval's basic argument is that in spite of R. Joseph Colon's prodigious use of the principle, he nowhere applies it to any of the great Ashkenazic scholars, such as the Maharil, the Mahari Weill, the Maharam of Fulda and others of that stature, who were active between the time of the Maharam and his own time. However, what Yuval fails to understand is that these authorities and others like them are hardly mentioned at all in any of R. Joseph Colon's responsa; R. Joseph Colon was simply not familiar with their writings. Although the Ashkenazic scholars, as a rule, made extensive use of the Maharik's responsa, the simple fact is that R. Joseph Colon was Italian, and not Ashkenazic, and was completely unfamiliar with the writings of Ashkenazic authorities later than the Maharam. The Maharik's principle is intended to cover "later" authorities in general, as he states:

> And also R. Meir, who was a later authority – we follow him ... and the reason is that the later authorities are more familiar with the explanations given by the earlier authorities than we are, yet in spite of this they still disagree with them ... likewise, on the basis of the same principle, we too must follow the later authorities ... it is our custom everywhere to follow the authorities who are later (*Responsa*, #181).

This approach was transmitted from R. Joseph Colon to R.M. Alashkar, and R.M. Alashkar's critical comments and ridicule target this very point. As for the Maharil and his colleagues, they could not have referred to any Ashkenazic scholar who lived after the middle of the fourteenth century, since the Maharil himself belonged to the first generation of scholars to emerge following the horrors of the Black Death and the other tribulations of that era.

According to Yuval, the principle that "the law is like the later" is only applied to Ashkenazic authorities in the period up to the middle of the fourteenth century, which is the historical boundary, the watershed, differentiating the periods in question. It is never applied to authorities who lived after the mid-fourteenth century, and certainly, never to even later, or contemporary, authorities. Indeed, he argues that even R. Moses Isserles did not apply the principle to anyone other than the Mordechai circle, the Rosh, and the Tur; the Rema's criticism of R. Joseph Caro is that he "did not adopt this decision-making rule," let alone give absolute precedence to this group of authorities over the Rambam and R. Isaac Alfasi. Yet Yuval himself quotes the words of R. Moses Isserles, who states, "It is known that the learned author of the *Beit Yosef* [R. Joseph Caro], had a natural inclination toward the great... Alfasi, Maimonides and Asheri... though they are early, not later authorities" (Yuval, *op cit.*; cf. the text above near note 5). Thus the Rema himself describes the Rosh as early!

Yuval's position is so extreme that he asserts that even R. Jacob Pollak and R. Shalom Shakhna said nothing more on the subject, and did not entertain the possibility of applying the principle to contemporary authorities. However, I must confess that I cannot comprehend his arguments for this.

A totally novel interpretation of the subject is offered by A. Reiner in his "Developments in the Yeshivas of Poland and Ashkenaz in the Sixteenth and Seventeenth Centuries and the Controversy over Polemics" (Hebrew), in *Jubilee Volume in Honor of H. Shmeruk* (Hebrew), (Jerusalem: 1993), p. 18. Further support for my position, generated in the discussion that arose following publication of my article, can be found in A.M. Rapeld, "The Scholars of Ashkenaz and Poland in the Fifteenth and Sixteenth Centuries on '*Hilkhata Kebatrai*'; Sources and Addenda" (Hebrew), *Sidra* 8 (1993), pp. 119–140.

I find incomprehensible Yuval's claim that "It is puzzling why this idea took 300 years (!) to penetrate the world of Jewish law and make an impact." After all, the idea was first introduced into the Jewish world around the middle of the thirteenth century; as described in my article, its dissemination and evolution took about 150 years; just what is it that Yuval finds so very puzzling about this?

There is no doubt that the changes in the way the principle "the law is like the later" was understood ultimately led to an increase in the decision-making power of contemporary authorities. This is true with respect to the way it was understood by all the Ashkenazic scholars, as is reflected in the treatment in the Shakh's "Klalei Horaat Isur Veheter," Yore Deia at the end of sec. 242; and as summarized in the *Encyclopedia Hatalmudit*. The only questions are why, when, and how these changes took place, questions I hope to have shed light on here.

CHAPTER 5

ON THE PLACE OF LOGIC (*SVARA*) IN MAIMONIDES' *CODE*

Shimshon Ettinger

1. Introduction
2. Rationales for Laws in the *Code*
   i. "Bringing the Laws Closer to Reason"
   ii. Rationales which Materially Impact on Rulings
   iii. Maimonides' Motivation for Citing Rationales; "Internal" and "External" Rationales
   iv. Classification of Maimonides' Rationales for Laws
3. Deciding between Different Sources
   i. "Redacted Talmud" and "Talmudic Debate"
   ii. Divorce due to Gentile Coercion
   iii. Logic as a Guiding Principle
   iv. Maimonides' Responsum on Invasion of Privacy
4. Maimonides' Attitude to Biblical Exegesis
   i. Biblical and Rabbinic Exegesis
   ii. Maimonides' Remarks on Exegesis in the *Book of Precepts*
   iii. Terminology Related to Exegesis
   iv. The Precepts of Prayer and Emulation of Divine Conduct
   v. The Source of the Injunction Against Perpetuating Dissent
   vi. The Disqualification of Women from Serving as Witnesses
5. Conclusion
6. Appendix: More on the *Code*'s Rationales for Laws

1. Introduction

This paper will examine Maimonides' use of reasoning, or logic, to aid in legal decision-making and the definitive formulation of the law.[1]

In Jewish jurisprudence, the concept of legal logic usually appears in the context of **interpreting** the sources;[2] as discussed here, however, "legal reasoning" will have a different sense. It will refer to logic as employed by Maimonides to analyze his sources: to decide whether a given source mandates its own use in determining the law, and if so, to demarcate its precise scope and status; and to select the determinative source in cases where there are a number of sources, conflicting sources, and so on.

Logic, in this sense, plays a role in the following three areas, which are examined below: (a) in providing rationales for the laws enumerated in the *Code*; (b) in choosing between different talmudic analyses of the same question; and (c) in determining Maimonides' attitude to laws established by the Sages on the basis of biblical exegesis.[3]

2. Rationales for Laws in the *Code*

i. "Bringing the Laws Closer to Reason"

It is widely known that Maimonides cited reasons and rationales of various sorts for many laws. In some cases, these rationales differ from those offered by the Talmud.[4] We fortunate enough to have the thoughts of Maimonides

---

1 Note that this sense of "legal reasoning" is distinct from that of "legal reasoning" as a source of the law, namely, the reasoning of the Sages. As is well known, the Sages used reasoning to arrive at many laws and legal principles. Cf. M. Elon, *Jewish Law* (Philadelphia and Jerusalem: 1994), vol. 2, pp. 987ff.

2 Obviously, in using the sources to arrive at legal conclusions, Maimonides must indeed interpret them as he sees fit. Commentators on Maimonides have for centuries been occupied – after identifying the sources on which Maimonides based his halakhic rulings – with the task of clarifying and reconstructing his interpretations of these sources.

3 Note that this does not refer to laws derived by means of the "thirteen exegetical modes" and other such methods, which Maimonides defines as "teachings of the scribes," but to laws inferred directly from interpretation of the Bible.

4 For more on this, see: R.B.Z. Benedict, "Ledarko Shel Harambam Betaamo Shel Hadin," in his *Harambam – Lelo Stiya Min Hatalmud* (Jerusalem: 1975), p. 47; J. Loewinger, *Darkei Hamahshava Hahilkhatit Shel Harambam* (Jerusalem: 1965), pp. 21ff., 92ff.; I. Twersky, *Introduction to the Code of Maimonides* (New

himself on this matter, directly expressed in a responsum concerning the Sages' injunction against the consumption of leavened bread "from the fifth hour" on the eve of Passover.

The Talmud implies that the injunction was issued to avoid transgression on a cloudy day, when people might become confused as to the time, and mistakenly believe the seventh hour, when the scriptural prohibition comes into force, to be the fifth.[5] Maimonides, however, states, "In the fifth hour leavened bread is not eaten; it is an injunction to prevent a mistake between the fifth hour and the sixth on a cloudy day."[6] The scholars of Lunel requested an explanation, and Maimonides replied:

> Alfasi says that one may mistake the fifth and seventh hours, and he is right. It is so stated by the Talmud.... But I mentioned the more common occurrence, which is substituting one hour for the next, because my sole intention in all my work is to bring the laws closer to reason or to what is most common.[7,8]

Haven: 1980), p. 162; and N.A. Rabinowitz, "Darko Shel Harambam Keposek Uparshan," in *Sefer Zikaron Lerav Yitzhak Nisim* (Jerusalem: 1975), p. 263. For further references, see the Appendix below.

5 bPesahim 12b; Alfasi *ad loc.* states this explicitly.

6 *Code*, Laws Concerning Leavened and Unleavened Bread 1:9.

7 I.e., to whatever is more commonplace. Cf.: "But I mentioned the more common occurrence ... because my sole intention in all my work is to come as close as possible to the ordinary occurrence" (*Kesef Mishne*, Laws Concerning Leavened and Unleavened Bread 1:9). In this context, note that elsewhere, too, Maimonides advocates another "principle of the most common," not in explanation of his own philosophy of the rationales for the laws, but as a general methodological principle of biblical law.

> You should also take note that the Law does not consider the curious and peculiar, the rare occurrence. Anything it wishes to teach us ... will be directed toward what is true in the majority of cases, and will disregard events which occur infrequently, or the harm which may befall some one individual due to a particular measure or policy of the Bible (*Guide for the Perplexed*, 3:34).

Cf. Maimonides' responsum on music (#224); and S. Rosenthal, "The Most Common," in *Prakim* 1 (Jerusalem: 1967–8), p. 183, as well as the articles by I. Englard and S. Rosenberg in *Shenaton Hamishpat Haivri* 14–15 (1988–89).

Here, Maimonides uses the principle to explain the fact that the Torah addresses ordinary people and behavior. A good illustration of how this explanatory principle is used in the *Code* is found in one of the laws of witnesses cited there, which

ii. Rationales which Materially Impact on Rulings

Mere departure from the rationales offered by the Talmud need not surprise us, given Maimonides' own explicit account of his intentions and approach to the matter. In essence, there is no problem as long as the rationale cited for a given law has no concrete repercussions – as long as no halakhic conclusions can be hermeneutically inferred from it, and it has no effect on the law's applicability and import in other cases. The rationale cited above for the injunction against eating leavened bread on Passover eve is of this kind, that is, it has no direct bearing on the law.

Discrepancies between the rationales for laws given in the Talmud and those cited by Maimonides are problematic only if they have consequences for the application of the law: that is, if the latter can be used as a basis for legal decisions that conflict with those that follow from the former.[9]

The fact is that even in such cases, Maimonides cites rationales that are at variance with those offered by the Talmud. What are we to make of this divergence?

---

conflicts with the same law in the Talmud: "The reason why a relative is unfit to testify, is not because we assume partiality, because he may testify neither for nor against a relative; rather, it is because this is a scriptural injunction" (Laws Concerning Evidence 13:16). The view that relatives are disqualified from testifying because of the scriptural injunction, and not because there is concern about the reliability of their testimony, is already found in the Talmud, but there, the grounds cited in its support are different.

> Perhaps the scriptural injunction that he is not believed but another is believed, is not due to prevarication: would Moses and Aaron be unfit to testify for their father-in-law because they are unreliable? It is a scriptural edict that they may not testify (bBaba Batra 159a).

Maimonides does not accept the reasoning of the Gemara with regard to Moses and Aaron, because they clearly do not represent the majority, and in deciding the law, curious unique exceptions are not taken into consideration. They would thus be disqualified from testifying even though the basic motivation for the law does not apply to them.

8 Maimonides, *Responsa*, Blau ed. (Jerusalem: 1960), #222; cf. *Kesef Mishne ad loc.*

9 Cf. the discussion of the ban on destruction of fruit trees during the sabbatical year in *Tosafot Yomtov* (Mishnah Sheviit 4:10). In the Talmud, this prohibition is based on scriptural exegesis: "Scripture [Lev. 25:6] says it should be eaten – not destroyed" (bPesahim 52b). In his *Commentary on the Mishnah*, however, Maimonides, offers a different reason: "because he robs others." (In the *Code*,

Let us first examine some examples.

(a) From the *Code*'s Laws Concerning Inheritance:

> One cannot bequeath to anyone not entitled to inherit, nor disinherit lawful heirs, though this is a monetary matter, for it is written in the pericope on inheritances, And it shall be to the children of Israel a statute of judgment [Num. 27:11], indicating that the law is immutable and not subject to qualification.[10]

The source of this law, according to which a testator cannot make any stipulations contravening the laws of inheritance, is the Mishnah, which states: "One who says, So-and-so, who is my first-born, shall not inherit a double

---

Laws Concerning the Sabbatical Year and the Year of the Jubilee 5:17, Maimonides does not cite this reason but quotes the Talmud; however, this is in keeping with the general policy of the *Code*, which is to reserve reasoned justifications for precepts based on the oral law, and not to offer them for biblical commandments; note that the reverse is true of the *Guide for the Perplexed*. There are many other examples, but space does not permit further discussion of the matter here. Let us simply note that we have no doubt that Maimonides is making an important statement on the relationship between the observance of a precept and its underlying rationale.) *Tosafot Yomtov* asks:

> I wonder why he took the step of explaining it by way of a different reason from that cited in the Gemara. ... and although the Torah can indeed be interpreted in many ways, this is the case when the interpretation is the end itself. But in deciding the law, we can rely only on the Sages of the Talmud, because there is no doubt that the law changes according to the rationale cited, as do analogies that are based on it.

It is interesting that *Tosafot Yomtov* offers a similar opinion on a related issue, though it is problematic. I refer to interpretation of the Mishnah in a manner at variance with that of the Gemara. In *Commentary on the Mishnah* (Nazir 5:5), *Tosafot Yomtov* quotes Maimonides' explanation and adds, "These are the words of Maimonides, and though the Talmud explains the matter otherwise, they are accepted, since they make no difference to the outcome of the law, hence it is permitted to offer an interpretation." See R.M. Roth's pamphlet *Simhat Yomtov*, printed as section 5 of his *Responsa Kol Mevaser* (Jerusalem: 1956). Roth takes issue with this view, arguing that it is difficult to determine, for any particular instance, whether or not the proffered interpretation affects the law's application. (Note that he cites the view of R.S. Luria as expressed in his *Hokhmat Shlomo*; Luria seems to concur with *Tosafot Yomtov*; R. Elijah of Vilna is reported to concur as well.)

10 6:1–2; cf. Benedict, *op. cit.*, p. 51.

portion, or, So-and-so, who is my son, shall not inherit with his brothers, his stipulation is void, because it contradicts what is said in the Torah."[11] The Talmud notes that this Mishnah disputes the view of R. Judah, who says that provisions contravening biblical law with respect to monetary matters are valid, and quotes a *baraita* which records an argument between R. Judah and R. Meir on the subject. "One who says to a woman: You are betrothed to me on condition that I am not bound by the usual marital duties [food, raiment and conjugal rights (Exod. 21:10)] ... even R. Judah says – in that case, she knew and waived her rights, in this case he does not waive his rights." In other words, according to the Talmud, the condition laid down by the testator is invalid, because the inheritor does not forgo his rights as does a woman who agrees to such a betrothal. This case is highly instructive, because here we have a situation where the Gemara brings a logical–legal rationale for the law, whereas Maimonides declares that it is a biblical injunction.[12]

Furthermore, another point crucial to our concerns emerges: at least one of the prevailing interpretations of the passage – that which is also the most feasible – contends that the dissenting rationale offered by Maimonides impacts on the application of the law. According to the commentary of R. Samuel b. Meir, the rationale in question, "in that case, she knew and waived her rights, in this case he does not waive his rights," should be understood as follows:

> The woman waived her rights and agreed to betrothal under the conditions indicated, because the betrothal was in her interest. But in the case of the will, who is there to waive any rights? How could it be in the interest of the son to forgo his inheritance? Even if the son was silent, this does not mean that he waived his right, rather, he did not want to antagonize his father.[13, 14]

On this understanding, it is conceivable that such a stipulation, namely, a stipulation to the effect that the testator disinherits his son, though incompatible with the laws of inheritance, might be valid, providing the son is informed of the clause and explicitly agrees to forgo his inheritance. According to Maimonides, however, such a clause cannot be valid, because a scriptural statute permits no exceptions.

---

11 Mishnah Baba Batra 8:5; bBaba Batra 126b.

12 In this regard it makes no difference whether it is true, as indeed it may be, that Maimonides was relying on a particular source for this scriptural exegesis of the term "statute" in Num. 27:11.

13 bBaba Batra 126b s.v. *hatam ka mahla*.

14 For more on this, see below.

(b) From the *Code*'s Laws Concerning Damage by Chattels:

> If an ox, innocuous or recidivate, of an Israelite, gores the ox of a non-Jew, the owner is exempt. Because the Gentiles do not make one liable for damages caused by his beast. So we judge them according to their law. But if an ox of a non-Jew gores the ox of an Israelite ... the owner must pay for all the damage. This is the penalty meted out to Gentiles, because they do not observe the precepts, and do not clear away obstacles, and if we do not make them liable for the damage caused by their animals, they will not control them and will cause others losses.[15]

This passage is of great interest, since it is obvious that here Maimonides differs from the Talmud in a way that has undeniable legal consequences. Scholars have wrestled with the passage for hundreds of years.

The Mishnah asserts:

> The ox of an Israelite which gores an ox belonging to the Sanctuary, or of the Sanctuary which gores an ox of an Israelite, is exempt, for Scripture states – the ox of his fellow – not the ox belonging to the Sanctuary. The ox of an Israelite which gores the ox of a Gentile is exempt, but that of a Gentile which gores the ox of an Israelite – innocuous or recidivate – the owner must pay for all the damage.[16]

The Gemara asks:

> We face a dilemma. If the scriptural restriction of the scope of this law to "his fellow" is accepted, then should not the Gentile whose ox gores that of the Israelite also be exempt? Conversely, if the restriction does not apply, then should not the Israelite whose ox gores that of a Gentile also be liable?

It then answers as follows:

> R. Abbahu cited this passage: He standeth and shaketh the earth; He beheldeth and maketh the nations to tremble [Hab. 3:6] – The Lord beheld that the seven commandments binding on the descendants of Noah were not observed ... . He declared them outside the protection of the civil law of Israel. R. Johanan said: He shined forth from Mount Paran [Deut. 33:2, where God offered the law to all the nations, but they declined to accept it]. At Paran He released their money to Israel.[17]

---

15 *Code*, Laws Concerning Damage by Chattels 8:5.

16 Mishnah Baba Kama 4:3; bBaba Kama 37b.

17 *Ibid.*, 28a.

First, we have to distinguish two distinct parts of the law as cited by Maimonides. In the first part, which refers to a case of an Israelite's ox goring that of a non-Jew, where the owner is exempt, he states, "because the Gentiles do not make one liable for damages. ... so we judge them according to their law." On this reasoning, which is not raised in the talmudic discussion, it immediately follows that should "their law," at some future time, recognize the liability accruing to animal owners for such damages, Jews would become liable for damages inflicted by their animals on non-Jewish animals. This conclusion cannot be reached on the basis of the reasons mentioned by the Talmud.[18]

---

18 The simplest explanation, accepted by most of the Early Authorities and by commentators on Maimonides, argues that the Talmud regards "his fellow" as an exclusive term, hence the Jewish owner of an animal which damages property of a non-Jew is exempt. R. Abbahu's exegesis, on this line of reasoning, is cited in reference to the second case, that of the ox of a Gentile goring that of an Israelite. Note that this reasoning differs from the argument suggested in the *Even Haazel* commentary on Maimonides, which, on the basis of *Tosafot R. Peretz*, suggests that according to Maimonides "his fellow" does not exclude non-Jews, but only animals that belong to the Sanctuary. Cf. *Shita Mekubetzet*, as well as note 22 below.

In any event, it seems certain that from R. Abbahu's rationale, we cannot infer that the law would change were the Gentiles to observe Noahide law, because R. Abbahu's statement is only a historical explanation for the biblical verse "beheldeth and maketh the nations to tremble," namely, that when the Almighty saw, at that time, that the seven Noahide commandments were not being observed, He penalized the nations and released their property to Israel. Given this fact, it is difficult to understand the explanation advanced by R. Menahem Hameiri, who may have been influenced by Maimonides: "It means that with respect to all those who observe Noahide law, legal liabilities are reciprocal."

R. Abraham b. David took issue with Maimonides: "This is not a sufficient reason, because the Gentiles do seize the damaging animal in lieu of damages." Apparently, he is suggesting that given this fact, the law should be changed accordingly. The *Magid Mishne* commentary says, in defense of Maimonides: "Perhaps during the talmudic period the Gentiles did not compensate for damages, and it may be that even today there are some Gentiles who do not do so. ... therefore even in a land where their law enjoins restitution, this law stands." This explanation is problematic and requires further explication. Cf. the *Hasdei David* commentary on the Tosefta, Baba Kama ch. 4, s.v. *shor shel goyim shehizik*; and *Even Haazel*. Both take issue with the view of Maimonides and *Magid Mishne*.

The *Migdal Oz* commentary[19] notes that the source of Maimonides' first rationale is the Jerusalem Talmud: "R. Abbahu says, in the name of R. Johanan – according to their law."[20, 21]

However, the matter is still not clear: first, why does Maimonides prefer the rationale given in the Jerusalem Talmud to that of the Babylonian, given that it clearly has repercussions for the application of the law?[22]

Second, the Jerusalem Talmud itself has reservations about the rationale of R. Johanan, which is the subject of a Tannaitic dispute, and, in the opinion of R. Judah the Prince, has been taken out of context. But a third difficulty is the most serious – the rationale in question does not explain the second clause of the law, which covers the case of a Gentile's ox which damages the ox of an Israelite. According to Maimonides, this calls for a "penalty meted out to Gentiles, because they do not observe the precepts." The significance of this rationale, as we will see below, extends well beyond what is discussed in the talmudic passage in question.

(c) From the *Code*'s Laws Concerning Idolatry:

> A town in the frontier regions is not condemned to destruction, in order to prevent idolaters from invading and destroying the land of Israel.[23]

The Talmud[24] attributes this rationale to R. Simon b. Yohai, whose mode of exegesis is generally to seek reasons for biblical passages. The Sages, however,

---

In his book on Maimonides, the Hazon Ish also disputes Maimonides' approach: "Perhaps the opinion of our master is that if they accept liability, we should do the same – to sanctify the holy name." This seems somewhat far-fetched, since Maimonides' words do not even hint at this. Cf. *Mirkevet Hamishne*, which, on finding Maimonides' stand problematic, turns to a discussion of the law's contemporary application, stating that since at present Gentile law mandates liability, should one of their oxen gore a Jew's ox, the owner is indeed exempt.

19 On the *Code, ad loc.*

20 jBaba Kama 4:3 (4b).

21 See the Jerusalem Talmud for other views on this question, which, as we noted, is also discussed in the Babylonian Talmud: "Rab said: He beheldeth and maketh the nations to tremble – he released the money of the Gentiles. Hezekiah said: He shined forth from Mount Paran … ."

22 Note that Maimonides himself accepts the standard exegesis of "his fellow" with regard to the law of double indemnity for theft (Laws Concerning Theft 2:1). This too contradicts R. Peretz, note 18 above.

23 *Code*, Laws Concerning Idolatry 4:4.

24 bSanhedrin 16b.

deduce this law from the wording of the relevant biblical verse: they interpret the expression "from the midst of thee"[25] as: "the Lord said from your midst and not from the far-off regions." Thus here too, Maimonides appends a rational explanation to a law, and is not satisfied with the exegesis of the Sages which renders the injunction a biblical statute.[26]

As noted, Maimonides prefers the controversial rationale of R. Simon b. Yohai to that of the majority of the Sages. The *Kesef Mishne* commentary explains:

> Since Maimonides rejects the method of R. Simon b. Yohai, as is evident from his view on the law of the widow's pledge, where he holds that a pledge is not taken from a rich woman,[27] he should not have given this reason. But because it makes no difference to the application of the law, and it brings the law closer to reason, and since the Sages do not deny it, but merely do not need it, for they have their own exegesis, Maimonides saw fit to mention it (*Kesef Mishne ad loc.*).

I find this explanation difficult to accept. It is apparent that both explanations have legal repercussions. It would be easy enough to find other means of preventing invasion by idolaters, and, were this to be done, according to R. Simon it would then be permissible to condemn a frontier town to destruction;[28] this, however, would not be the case were we to reject this rationale and uphold only the fact that it is a biblical injunction. [29, 30]

---

25 Deut. 13:14–15.

26 This is also true of many other laws in the *Code*; see the Appendix, below.

27
> The Sages taught: No pledge is taken from a widow, whether she be rich or poor, so says R. Judah. R. Simon says: If she is rich a pledge is taken, if she is poor, no pledge is taken, for you must return it, and this may mar her good name (bBaba Metzia 115a).

Maimonides cites the law according to R. Judah, and the Mishnah (*Code*, Laws Concerning Creditor and Debtor 3:1).

28 After all, the phrase "to interpret the reason behind the biblical verse," attributed by the Talmud to R. Simon, means to take into account the rationale underlying the law in order to draw operative conclusions, and not just to provide an explanation. Cf. the passage on the widow's pledge.

29 There is a further example, one which is not quite unequivocal. In the *Code*'s Laws Concerning Circumcision, Maimonides writes, "An infant is only circumcised if

iii. Maimonides' Motivation for Citing Rationales; "Internal" and "External" Rationales

In view of the laws discussed above, characterized in each case by the fact that Maimonides explained the reason for the law in a manner at variance with that of the Talmud, even though doing so gave rise to material differences

---

he does not suffer from any illness, because mortal danger overrides everything. It is possible to circumcise after the usual time, but a life ... can never be returned"(Laws Concerning Circumcision 1:8).

The second reason, "it is possible...," is not mentioned by the Talmud, and could be interpreted sophistically as saying that if circumcision cannot take place at a later date, it must be performed at once, even if this gives rise to mortal danger! Cf. Hatam Sofer, *Responsa*, Yore Deia #245; Benedict, *op. cit.*, p. 55.

We do not suggest that this was indeed Maimonides's intention, because he himself says: "mortal danger overrides everything." Nevertheless, the reasons for this second rationale are unclear, and it is hard to understand why Maimonides did not omit this problematic rationale, given that such an inane conclusion can arguably be derived from it.

30 Loewinger (*op. cit.*, p. 70) makes an interesting point: Though Maimonides rules against R. Simon regarding the law of the widow's pledge – against, so it would seem in light of the talmudic discussion, his policy of "interpreting the reason behind the biblical verse" – he nevertheless rules according to R. Simon elsewhere. This is so even though the Gemara states that the very same method is used there – i.e., that R. Simon "interprets the reason behind the verse." The passage in question concerns the prohibition against intermarriage with non-Jews. In the *Code*, Maimonides states:

> An Israelite cohabiting with a non-Jewish woman by way of legal marriage ... they are flogged, according to Scriptural law, for it is written "do not marry them"; this prohibition includes the seven nations and all other nationalities (Laws Concerning Forbidden Intercourse 12:1).

In my opinion, this example does not support Loewinger, who contends that Maimonides ignores a fundamental connection between the two issues, which the Talmud upholds. (Loewinger brings other examples in support of this claim.) In the case before us, it seems, Maimonides disregards the association because of the way in which he interprets the following passage from bKidushin:

> How do we know that intermarriage is forbidden? The Bible tells us: "neither shall you make marriage with them." ... Now that verse refers to the seven nations – how do we know it applies to other nations? The Bible states: "For he will turn away your son" [Deut. 7:4] – this includes all who

with respect to the application of the laws in question, we can assume that Maimonides did not intend these reasons to serve as a basis for further interpretation and for drawing legal conclusions. The purpose of these explanations is, as Maimonides himself attests in his responsum, "to bring the laws closer to reason," nothing else. They cannot teach us the specifics of the law or serve as a source for extrapolation to other cases.

However, we should note that this is not universally true. In my opinion, we must take into consideration the familiar distinction, found in both Jewish law and law in general, between external and internal rationales.[31]

> would turn him away. This is sufficient for R. Simon, who interprets the reason behind the verse, but what reason is given by the Sages for the prohibition? (bKidushin 68b)

According to Rashi (who also interprets the passage in bJebamot 23a in this manner), the Gemara's intention is to make an inference on the basis of the "surplus" language of the biblical passage; according to R. Simon, who interprets the reason for the verse: "there is no need for me to interpret the reason behind the verse, consequently, 'For he will turn away your son,' is redundant, and must therefore indicate that other nations are included in the prohibition." This is not the case on the reasoning of the Sages, who do not, themselves, interpret the reason for the verse; for them, "for he will turn away" is not redundant.

Maimonides, apparently, interprets this passage along the lines suggested by R. Nissim Gerondi (on *Alfasi ad loc.*, Vilna ed., p. 29b). On this interpretation, R. Simon does not apply the prohibition on intermarriage to all nations on the basis of exegesis but on the strength of the reason manifest in the words of the verse, "For he will turn away your son from following me." This clearly indicates that the prohibition does not apply only to the seven nations referred to, but to any nation that arouses the same fear, namely, that it will turn the Israelites away from the Almighty to idol-worship.

Another case like this, where the reason for the injunction is clearly expressed in the text itself, is the discussion of the widow's pledge (Baba Metzia 116a). Even R. Judah, who takes issue with R. Simon's method – interpreting the reasons behind biblical verses, agrees that if the reason is explicitly stated in the biblical text itself, we are to draw legal conclusions from it. Thus while there is controversy in the Talmud (though from the passage in bKidushin, we can see that the Sages differed with R. Simon even on this), there is little doubt that, on logical grounds, Maimonides favored the position taken in the passage in bBaba Metzia over that expressed in the discussion in bKidushin. See below, sec. 3; cf. *Shaar Hamelekh* on Laws Concerning Forbidden Intercourse *ad loc.*, and *Lehem Mishne* on Laws Concerning Creditor and Debtor 3:1.

31 See Y. Kuperman, *Lipshuto Shel Mikra* (Jerusalem: 1974), p. 80, note 32; D. Hanschke, "Ein Mikra Yotzei Midei Pshuto" *Hamaayan* 17 (1977), p. 18.

An internal rationale in effect specifies the **legal parameters** of the law, and is, therefore, an integral part of it. The substance of an external rationale, on the other hand, is either some sort of external objective, whether moral, socio-economic, or other, which the law in question hopes to achieve; or an **external factor**, such as a historical consideration, which determines the nature of the law. Some examples will make this clearer.

An example of an internal rationale in Jewish law is the law of "interested witnesses."[32] The reason for the disqualification of interested witnesses from testifying in court is a contentious issue for both the Early and the Later Authorities. The obvious explanation is the fear of false testimony prompted by the witness's interest in the outcome of the case; in the opinion of some authorities, this is indeed the rationale. Others, however, deny this, arguing instead that the interest turns such a witness into "one who testifies about himself" – a relative, as it were. Now relatives are barred from giving evidence, and it is unanimously accepted that the reason for this prohibition is not doubt as to their reliability, but the "scriptural injunction" against it.[33]

It is easy to appreciate that the rationale behind this law, according to both opinions, has no external purpose, but is an integral part of it; it demarcates the law itself. Halakhic scholars agree that rationales of this kind affect legal rulings. Indeed, this is tautologous, since, as we said, such rationales delineate the laws in question and cannot be separated from them. In this example, the fact that the legal outcome could vary according to the rationale adopted would be manifest were the interested witness to testify **against** his own interest. On the first opinion, that is, on the view that such a witness may perjure himself to advance his self-interest, such apprehension is clearly misplaced here. But if it is maintained that the interested witness is disqualified because he is a "related party," he cannot testify, whether for or against himself. And in fact, the authorities who disagree on the rationale also disagree on whether an interested witness can testify against himself.[34, 35]

---

32 *Code*, Laws Concerning Evidence 13:15.

33 *Code*, *ibid.*

34 "One who sells a field to his fellow without a guarantee does not testify as to its status" (bBaba Batra 44b); and see ibn Migash *ad loc.* Cf. "One who sold his fellow a house or a field does not testify as to its status" (bBaba Batra 43b). For commentary on this see *Nimukei Yosef* on Alfasi *ad loc.*; see the *Shakh* and *Sema* commentaries on Hoshen Mishpat 37:1; and *Ketzot Hahoshen* 37:9.

35 Another example of an internal rationale is the law "no one can be considered an agent in carrying out a transgression." The Talmud gives the following rationale

On the other hand, if the rationale is external, that is, if it has an external aim[36] – then it is not an integral part of the law itself, and it may well be that we should neither interpret it nor infer any halakhic conclusions from it. Although as a rule the identification and classification of the rationale for a law is relatively uncomplicated, some cases are more problematic.[37]

iv. Classification of Maimonides' Rationales for Laws

Three of Maimonides' four laws discussed above[38] have only external rationales, and our assumption is that Maimonides did not intend these rationales to carry any legal weight.

The rationale for the law, "A town in the frontier regions is not condemned," namely, prevention of invasion by foreigners who would devastate the land, is undoubtedly external – it has a clear outward aim: to secure the borders of the land of Israel. The rationale for the law concerning damage wrought by animals of Israelites on those of Gentiles, and vice versa, is similarly external, to wit, "because the Gentiles do not make a man liable for the

---

for this law: "A teacher's statement and that of his student – to whom do we listen?" (bKidushin 42b). That is, the principal is entitled to assume that his agent will not carry out a transgression. However, there are other accounts of the reason for the law, and in some cases, there could be legal consequences if one rationale is adopted rather than another; for example, consider a case where the agent has previously been guilty of the transgression in question. Furthermore, there would clearly be a different legal outcome if we accept the other reason cited in the Talmud, namely, that it is a biblical injunction. Cf. *Encyclopedia Talmudit*, s.v. *ein shaliah ledvar aveira.*

36 For examples, see the next section.

37 Of course, I do not mean to imply that every law has both an internal and an external rationale. The examples presented illustrate this. Further, there is another argument for the distinction between external and internal rationales. All the authorities decide against R. Simon, whose method of "interpreting the reason behind the verse" they reject (see note 30 above); yet on several occasions they themselves do not hesitate to offer exegetical interpretations, from which they proceed to draw legal conclusions. The solution to this paradox is to note that R. Simon's method of "interpreting the reason behind the verse" is applied only when the rationale in question is **external,** as is clear from the case of the widow's pledge and other cases where the Sages disagree with him; later scholars are also opposed to his interpretaion. However, they certainly do not object to such interpretation when the rationale is internal, when it defines the parameters of the law. In such cases, it is plain that legal conclusions should indeed be drawn.

38 Including the example in note 30.

damages caused by his beast," and further, as a penalty imposed on the Gentiles for not preventing their animals from doing damage. Here too, it is clear that the rationale is not an integral part of the law, a delineation of the law's parameters, but an external aim or consideration that motivates it.[39]

The exception is the law dealing with testamentary conditions contravening the biblical laws of inheritance.[40] The Talmud distinguishes between conditions qualifying betrothals and those qualifying inheritance: "in that case she knew and waived her rights, in this case he does not waive his rights." The testator cannot qualify the laws of inheritance because, unlike the bride who agrees to forgo her marital rights, the beneficiary does not waive his right to his inheritance.

There is no doubt that the element of waiving or not waiving one's rights is not a goal or an external cause, but an internal rationale demarcating the parameters of the law in question. It is thus noteworthy that Maimonides does not cite it, but instead chooses to explain the law as a biblical precept.

However, it is not difficult to understand why Maimonides does not use the internal rationale which is cited in the Talmud. Further scrutiny of the sources reveals that this rationale deviates from the plain sense of the Mishnah, both in terms of its language, and in terms of its logic. The Mishnah says, "One who says, So-and-so, who is my first-born, shall not inherit a double portion ... has said nothing, because his provision contradicts what is said in the Torah." This implies that the provision is ineffective under any circumstances. But the explanation in the Gemara, "in this case he does not waive his rights," certainly seems to imply, as R. Samuel b. Meir explains,[41] that if the son explicitly renounces his rights, the provision is valid.[42]

---

39 It is difficult to articulate precisely the rationale for delaying the circumcision of an unhealthy child – "it is possible to circumcise after the allotted time" (see note 29). We feel, nevertheless, that this is also a purely external rationale, the import of which is the weighing of conflicting considerations and selection of the most reasonable, all the more so since immediately before this point Maimonides cites the principle that "mortal danger overrules everything."

40 Example (a) above.

41 Also see Hameiri's *Beit Habehira* on Baba Batra 126b (Sofer ed., p. 517). For an explanation of the Talmud's stance on the matter, see S. Albeck, *Dinei Hamamonot Batalmud* (Tel-Aviv: 1976), p. 341.

42 The early commentators had some difficulty explaining the Gemara's position. Some felt obliged to say that in spite of what the text appears to imply, the son's waiver would not be effective. See, for example, *Yad Rama* on Baba Batra *ad loc.*

That is, because the rationale in question is internal, stipulating the legal limits of the law, Maimonides correctly inferred that it was legitimate for him to offer an interpretation and apply the law in accordance with it. But unfortunately, in this case, his conclusion conflicts with the plain meaning of the Mishnah.[43]

It appears, then, that Maimonides intended to restore the Mishnaic law, and hence decided to deviate from the rationale cited in the Gemara and cite the law's status as a biblical injunction: "for it is written in the chapter on inheritances: 'and it shall be to the children of Israel a statute of judgment.'"[44]

3. Deciding Between Different Sources

What part does logic play in determining which sources will be used to decide the law, given that different sources may generate conflicting decisions?[45] We now turn to an examination of this question. Most of the sources in question are in the Talmud.

---

s.v. *beram tzarikh et lemeida*: "with regard to the son's inheritance – the son was entitled to it from the moment of his birth. He surely did not relinquish his rights then." Cf. *Hidushei Harashba ad loc.*: "even if he did make a waiver, it waived nothing and is void"; and compare *Hidushei Haran* and R. Yomtov of Seville *ad loc.*

43 There may be another argument against the rationale proffered by the Talmud. Inheritance is a one-sided transaction, and the role of the beneficiary is purely passive, hence his waiver is useless and meaningless. But this argument has its flaws – see Albeck, *op. cit.*, (note 41 above), p. 342. Note that the matter of the waiver is cited by the Gemara as a rejoinder raised solely to harmonize the Mishnah wlth the opinion of R. Judah regarding "one who makes a stipulation contravening a biblical injunction." Though it is, of course, possible that the Mishnah follows R. Meir; see below.

44 Whether Maimonides himself drew halakhic conclusions from the rationales cited in the Talmud is a different question, requiring its own investigation. At present I have seen no evidence that he in fact did.

45 This is relevant to legal decision-making in general. Both the Talmud and the Geonim articulate rules for deciding between conflicting opinions in the Talmud, controversies between the Jerusalem Talmud and the Babylonian, etc. It is interesting that Maimonides does not always follow these rules in his own decision-making, a fact that calls for explanation. Though this article focuses mainly on conflicting sources rather than on resolving disputes between the Sages, the two subjects invite comparison and inference. See notes 71 and 97 below.

i. "Redacted Talmud" and "Talmudic Debate"

A key question is that of Maimonides' conception of the role played by talmudic debate in determining the law. This issue was addressed by Maimonides himself, and of course, his comments are of singular importance. They were made in the context of a question regarding the laws governing forbidden sexual relations. Maimonides was asked to clarify his decision with regard to the flogging incurred by one who was guilty of intercourse with a female bastard (*mamzer*), or other actions contravening negative scriptural injunctions (*hayavei lavin*).

> Whether a male bastard marries an Israelite woman, or an Israelite marries a female bastard, they are liable to a flogging if intercourse follows betrothal. If he betroths her but does not have intercourse with her, he does not incur a flogging. If he has intercourse without betrothing her, he is not liable to a flogging on account of bastardy. The only case in which a man is liable to a flogging for intercourse without betrothal within one of the unions forbidden by a negative commandment, is that of a High Priest who has intercourse with a widow, as will be explained.[46]

A review of the Talmud's account of the matter,[47] however, reveals that Maimonides' ruling is not in harmony with it. The Talmud discusses whether one who is liable to a flogging must also pay a fine. Since, elsewhere in the Mishnah (Mishnah Makot), it is stated that one who is flogged does not pay, why does the Mishnah here say that "these are the maidens intercourse with whom incurs a fine" and proceed to list maidens of whom it is said that intercourse with them incurs a flogging?[48]

In response, the Gemara brings the opinion of Resh Lakish: "This is the view of R. Meir, who says he is flogged and fined."[49] The opinion of Rabba is then cited. Rabba contends that according to R. Meir, "In demanding a fine the Torah is creating an exceptional case – that although he is to be put to death, he must pay a fine."[50] R. Pappa finds this problematic:

46 *Code*, Laws Concerning Forbidden Intercourse 15:2. The term "*mamzer*" denotes a child born of an incestuous relationship or from adultery involving a married woman.

47 bKetubot 3lbff.

48 Such as one's sister and a female bastard. See *Tosafot ad loc.* s.v. *vereminhu.*

49 bKetubot 33b.

50 *Ibid.*, 35b.

> R. Pappa said to Abbaye about it, Rabba said that the Torah was creating an exceptional case in demanding the fine; in accordance with whom is the Mishnah? If it is in accordance with R. Meir, the case of incest with a daughter becomes a problem, if it is in accordance with R. Nehuniya b. Hakane, that of incest with a sister becomes a problem, if it is in accordance with R. Isaac, that of the female bastard becomes a problem.

As Rabba explains the opinion of R. Meir, this Mishnah is not in agreement with any of the Tannaitic opinions on the matter. The Talmud concludes "of necessity it is according to the opinion of R. Johanan"; according to R. Johanan, the Mishnah is referring to offenders who were not forewarned about the impermissibility of the act and hence cannot be flogged; since they are not flogged they can be fined.

The implication is that one who has intercourse with a bastard after being forewarned is flogged; while Maimonides clearly rules that one is only flogged for such intercourse if he betrothed the woman first. Now the rapist of a female bastard certainly does not betroth his victim before the act. Indeed, R. Abraham b. David critically comments: "This is obscure, and not in harmony with the other cases."

Maimonides was asked by the scholars of Lunel:

> It is also written in your book that a man who has intercourse forbidden by a negative commandment is not liable to a flogging unless he betrothed his partner. We searched diligently but could not discover your source. But we do have great difficulty in explaining the list [in Mishnah Ketubot] given your opinion.[51]

Maimonides replies at length: "This question and your critical point are well taken and require much study." He explains that originally he intended to rule in accordance with the view of the passage in Ketubot as expressed by R. Isaac, but in the end, "after having copied and examined all the different opinions in the various books, I decided to overlook the difficulty, because I said, **let us not abandon redacted Talmud** (*talmud arukh*) **in favor of a talmudic debate** (*masa umatan shel gemara*)" [here, and throughout, emphasis in quotations from sources is added].

The phrase "redacted Talmud" here, which is contrasted with the talmudic debate in the passage we have been examining, refers to the discussion of the matter in bKidushin 78a. The Talmud there quotes an argument between Abbaye and Rava: "Abbaye says, he [the High Priest] betroths – he is

51 Maimonides, *Responsa*, Blau ed., #345.

flogged, because she is forbidden; he had intercourse – he is flogged for having defiled his seed. Rava says: he is flogged only if he has intercourse with her. If he does not, he is not flogged." According to Rava, then, in accordance with whom the law is decided, a flogging is only incurred if the High Priest, after betrothing the woman, had intercourse with her. At the end of his answer, Maimonides reiterates: "In any case we do not decide the law on the basis of a talmudic debate, as I said, and I believe there is no doubt at all about this."

We can learn an important lesson from Maimonides here, one which touches directly on the question of his policy with regard to deciding between different talmudic sources: "redacted Talmud" is to be preferred to "talmudic debate." The former term, denoting a clear, decided law, appears a number of times in the Talmud.[52] There is even an occurrence of the term in a context very similar to that of the present discussion, and it is possible that Maimonides intends the term to be understood here as it is there. Moreover, examination of that passage will also shed light on the contrasting term, "talmudic debate." The Talmud states, "This Mishnah is in accordance with Admon, and what I say is not just an ad hoc argument, but redacted Talmud."[53]

In this passage, the term "redacted Talmud" is contrasted with "an ad hoc argument." Clearly, "redacted Talmud" refers to an established teaching or interpretation, as opposed to an ad hoc suggestion for resolving an immediate problem or refuting an argument taken from a different context. Thus Maimonides' "talmudic debate" is much like the Talmud's "ad hoc argument," for it too connotes remarks made in the context of an academic discussion about legal matters, undertaken to rebut a dissenting opinion or resolve a contradiction between different sources. Indeed, these very features characterize both of the passages which Maimonides refers to as "talmudic

52 Cf. bZebahim 9a: "we have redacted Talmud that phylacteries are considered to be barriers"; and bMenahot 66b. Maimonides himself also uses the term in his famous epistle to R. Pinhas the Judge of Alexandria: "know that all that is unattributed is based on redacted Talmud – the Babylonian Talmud or the Jerusalem Talmud, or Sifra or Sifre or the redacted Mishnah or the Tosefta ..." [*Letters and Responsa* (Lipsia: 1859), vol. 1, sec. 140, p. 26]. It thus seems that the term as used here is close to its meaning in the talmudic passages cited above. Cf. Twersky, *op. cit.* (note 4), p. 157.

53 bShevuot 40b, the case of someone sued for a debt of wheat, who acknowledges a debt of barley.

debate": on the one hand, the conclusion of the passage from Ketubot arises out of the problem R. Pappa posed to Abbaye regarding Rava's opinion: "In accordance with whom is the Mishnah?... if according to R. Isaac, the case of the female bastard becomes a problem." On the other, we have Rava's explicit statement, which speaks for itself, in the passage from bKidushin.

Even assuming that we are correct as to the significance of Maimonides' principle for resolving conflicts between sources, two further questions must be answered. First, we must find out how the principle is to be applied in practice in other cases, and more importantly, we must see if this principle for deciding between incompatible sources suffices, as it stands, to provide a satisfactory resolution of all the problematic passages we have encountered, especially those where it is difficult to understand the logic underlying Maimonides' choices between such sources.

ii. Divorce due to Gentile Coercion

Another law in the *Code* the relationship of which to the Talmud's position has been debated by commentators and scholars is that governing the validity of a divorce due to Gentile coercion.[54] The Mishnah says: "A divorce coerced by Jews is valid, a divorce due to Gentile coercion is invalid."[55] The Gemara comments: "R. Nahman said in the name of Samuel: if it was legally coerced by Jews, it is valid, if illegally, it is invalid; if it is coerced legally by non-Jews, it is invalid, illegally by non-Jews – it is not even close to being a divorce."

The question is posed: "But how can this be? If the non-Jews are allowed to exert coercion, it should be valid, etc." This elicits the following response from R. Mesharshia: "According to the Torah, a divorce coerced by non-Jews is valid, so what would be the reason for saying it is invalid? To prevent every Jewish woman from going and finding a non-Jew and thereby releasing herself from her husband." That is, the Sages issued an enactment to the effect that divorces coerced by Gentiles are invalid to keep Jewish women from using this as a pretext for resorting to non-Jewish courts. The Talmud remains unconvinced and concludes: "The suggestion of R. Mesharshia is but an empty illusion."

---

54 Cf. Benedict, *op. cit.*, pp. 145ff; Loewinger, *op. cit.*, p. 164. The index on page 580 of the Frankel edition of the *Code*, Book of Women, lists the authorities who discuss this problem.

55 bGitin 88b.

But surprisingly, Maimonides rules against the Talmud, favoring the opinion of R. Mesharshia:

> Regarding one who is legally obliged to divorce his wife but refuses to do so – it is within the authority of any Jewish court, wherever and whenever convened, to flog him until he acquiesces and writes a bill of divorce, which is valid. Similarly, if he was flogged by non-Jews, who told him, "Do as you are instructed by the Jews"... if the non-Jews themselves compelled him to write the bill of divorce, since he is legally obliged, the bill of divorce is invalid (*pasul*).[56, 57]

The commentators[58] point to another passage in the Talmud in which the opinion of R. Mesharshia is upheld.[59] Its subject is sale under duress. R. Huna's ruling is: "If he was left suspended [tortured] until he agreed to sell, the sale is valid." The Gemara asks: "For what reason?" and, after a number of tentative replies, finally answers, "The argument is that due to the coercion the vendor capitulated and sold." If so, asks the Talmud, "R. Judah points out a difficulty: a divorce coerced by Jews is valid but by Gentiles invalid – why? In those circumstances too we can say due to the coercion he capitulated and divorced her." The Talmud's answer is as follows:

> But has it not been stated in this regard, "R. Mesharshia said, According to the Torah, a divorce forced by non-Jews is valid, so what would be the reason for saying it is invalid? To prevent every Jewish woman from going and finding...."

The opinion attributed to R. Mesharshia is not rejected.

The question is whether this passage can indeed be the source for Maimonides' ruling. Is the relationship between the passage in bGitin and that in bBaba Batra that of "redacted Talmud" to "talmudic debate"? It could be argued that the rejection of R. Mesharshia's opinion in bGitin takes place at the conclusion of a "talmudic debate" which seeks to reconcile his view with that of R. Nahman, who cites Mar Samuel. However, it is not as easy to argue that the passage on forced sale is "redacted." Closer scrutiny reveals that it is not; in fact, the reverse is true. The discussion in bBaba Batra is clearly far from "redacted," given that the statement of R. Mesharshia is introduced only to resolve a contradiction noted there between the argument

56 *Code*, Laws Concerning Divorce 2:20.
57 That is, invalid by decree of the Sages; see *ibid.*, law 7.
58 *Magid Mishne* and *Kesef Mishne ad loc.*
59 bBaba Batra 48a.

put forward to explain the ruling on sale under duress and the law of coerced divorce. Yet though it seems that in both places we are dealing with "talmudic debate," as opposed to "redacted Talmud," the passage from bGitin is still more definitive than that from bBaba bBatra. Indeed, the latter can hardly be said to constitute a discussion of the question of coerced divorce at all, let alone a definitive resolution of the matter: the issue of coerced divorce only comes up **incidentally** in the context of attempting to resolve a contradiction in the explanation of the forced sale ruling.

The passage in bGitin, on the other hand, is a focused, detailed and comprehensive discussion that examines the question raised in the Mishnah about divorce due to coercion by Jews or by Gentiles. On what grounds, then, could the incidental treatment of the subject in bBaba Batra be considered more definitive than the considered account in bGitin?[60]

iii. Logic as a Guiding Principle

It is important, I believe, to consider another major element in Maimonides' decision-making, especially in the context we are examining, that of deciding between conflicting sources. I refer to the role of logic.

To explore this issue, we will briefly return to the passage considered above on the flogging of those guilty of forbidden intercourse. It will be recalled that Maimonides ruled, in opposition to the Talmud, that one is not flogged for intercourse contravening negative scriptural injunctions unless

---

60 Cf. the view of R. Solomon b. Simon Duran:

> But in Laws Concerning Divorce, Maimonides wrote this. ... and it seems to me that in so writing he relied on the passage about the forced sale: If he was left suspended [tortured] until he agreed to sell ... . It is strange, however, that in the passage discussing the coerced divorce, the view of R. Mesharshia is rejected. ... if only I could understand it and support it, and I have not seen anyone contest this (Rashbash, *Responsa*, #339, p. 67a).

Also see *Kesef Mishne*, which also refers to bBaba Batra, as we saw: "It is hard to understand why we should rely on a remark which is regarded as an 'empty illusion.'"

See Benedict, *op. cit.*, p. 157, on the principle that a passage which occurs in its logical context (*sugia bimkoma*) is more authoritative than one which does not. In my opinion, this analysis does not apply to the passages discussed here, since, as I explained above, the remarks in bBaba Batra are made incidentally.

one betroths the woman beforehand. In a responsum, he explained that he favored the "redacted Talmud" of the passage in bKidushin over "talmudic debate." Upon closer examination, however, we find that in addition to this formal reason, Maimonides also endorsed Rava's ruling in bKidushin because it manifestly makes more sense.

Note that the law in question has to do with conduct forbidden by the Bible: "A bastard shall not enter into the assembly of the Lord."[61] The act in question – the rape of a female bastard – surely does not constitute an instance of a bastard's "entering the assembly." For the Bible is undoubtedly referring to becoming a member of the community through marriage. This was pointed out by none other than R. Abraham the son of Maimonides, who, responding to critical comments, stated:

> See how fastidious my father was in his work; with regard to the prohibitions pertaining to priests, the Bible does not use the phrase "shall not enter the assembly of the Lord," which it uses in stipulating the other negative injunctions; "entry into the assembly" can only be achieved by betrothal, whereas promiscuous fornication cannot constitute "entry into the assembly."[62]

Having clarified this, it will be useful to reconsider the law governing divorce due to coercion by non-Jews. This law has to do with cases in which, by law, the man is to be coerced into giving his wife the writ of divorce,[63] but the coercion is carried out by non-Jews. Is it logical to consider this divorce invalid **in principle** as a matter of biblical law, rather than as a matter of Rabbinic disqualification, only because it was a non-Jew who coerced the husband to discharge his obligation? For after all, the man was legally obliged to do so, and what have the characteristics of the agent who brought it about to do with the **substance** of the divorce?[64] Apparently, this is the reason Maimonides adopts the view of R. Mesharshia, in spite of its rejection by the Talmud.

---

61 Deut. 23:3.

62 *Birkat Abraham* #43; cf. *Kesef Mishne*, Laws Concerning Forbidden Intercourse 15:2.

63 Rashi, Gitin *ad loc.* s.v. *kedin*: "such as those – that is, cases in which he must divorce her and must pay her *ketuba*, or she was forbidden to him"; "one who is legally obliged to divorce his wife but refuses to do so" (*Code*, Laws Concerning Divorce 2:20).

64 Note the wording of Maimonides' law: "…or the non-Jews themselves compelled him to write the writ of divorce, since he is legally obliged, the writ of divorce is invalid" (*Code*, *ibid.*).

iv. Maimonides' Responsum on Invasion of Privacy

The centrality of logic in the decision-making process is clearly expressed by Maimonides himself in a well-known responsum concerning the invasion of privacy in a garden. The first Mishnah in bBaba Batra states: "Regarding those with a common yard who wanted to erect a partition – the wall is built in the middle. ... and also in a [common] garden."

The Gemara records two variants of the law: damage by seeing [into a neighbor's yard] constitutes invasion of privacy; and, damage by seeing does not constitute invasion of privacy.[65] One of the arguments the Gemara brings against the latter opinion is based on the Mishnah's ruling regarding the garden: "and also in a garden." The response to this is,

> A garden is different, like R. Abba said, as quoted by R. Huna, who said, in the name of Rav: it is forbidden to gaze at one's fellow's field when it is laden with crops.

According to the passage, there is no doubt that in a garden, to be exposed to one's neighbors' glance is to incur damage, and thus, by law, a wall should indeed be built, as in a courtyard. Maimonides, however, in upholding the rule that damage by seeing does indeed constitute invasion of privacy, rules against the Talmud regarding the law as applied to gardens: "and also a garden, he may insist on its separation from his neighbor by a wall ten handbreadths [three feet] high."[66] R. Abraham b. David of Posquieres notes, however, "This is confused, the wall should be four cubits [six feet] high [to prevent damage by seeing]" (*ad loc.*).

Maimonides was asked about this too by the scholars of Lunel, and answered them as follows:

> To ask a question such as this is not appropriate for great scholars such as yourselves. Do you not differentiate then between "damage by sight," which is certainly substantial damage – for one to see his neighbor standing, sitting or going about his business, and the other kind of "damage by sight," which is looking at another's field full of standing crops and possibly harming it by the evil eye? The latter is a supererogatory consideration only, to prevent neighbors from inviting the evil eye, and it is an incidental remark (*shinuyia bealma*) in the Talmud, which is not an integral part of the law. The real damage is in a courtyard where

65 bBaba Batra 2a–b.
66 *Code*, Laws Concerning Neighbors 2:16.

> people actually live. ... if we were to take the element of evil eye seriously, then the erection of walls could be made compulsory even in the open spaces where most of the fields are.[67]

Maimonides' statement that the argument in the Gemara, "a garden is different," is merely an incidental remark and not part of the law, comes as no surprise. Although Maimonides could have arrived at this conclusion independently, on the basis of his own reasoning, it so happens that in this case some of the Geonim had reached this conclusion before him.[68] In the words of R. Hananel,

> And the first version that states that damage by seeing does not constitute invasion of privacy, is not the law. For they raised six difficulties about it ... and the answers to all of them are unauthoritative, they are simply rejected and not to be relied on, and this is not controversial.[69]

This does not mitigate our surprise at Maimonides' ruling, since the argument "a garden is different" implies that regarding the garden everyone agrees that there is an invasion of privacy, even those who maintain that damage by seeing does not constitute an invasion of privacy. For those who hold that damage by seeing is indeed an invasion of privacy, there is absolutely no doubt that this holds in a garden too. This follows even from the Gemara's remark: "Come and hear, and also in a garden."

But as we said, Maimonides makes his own view clear: "Do you not differentiate, then, between damage by sight which is certainly substantial damage ... and the other kind of damage by sight, which is looking at another's field full of crops and possibly harming it by the evil eye?" The important consideration for him here is logical reasoning, upon which he bases this distinction between the substantial damage incurred when a courtyard is involved, and the minor damage in the case of a garden.

Nonetheless, Maimonides does not let his case rest on this alone, and he further bolsters it by citing another law in the same talmudic passage, that

---

67 Maimonides, *Responsa*, #395; cf. *Magid Mishne ad loc.*

68 It can be proved that the terms "strained reply" and "incidental remark" as used by early scholars usually refer to conclusions based on arguments from logic. Cf. H. Albeck, *Mavo Latalmudim* (Tel Aviv: 1969), pp. 545, 549; Y. Dinari, "Hageonim Harishonim Ubeayat Hashinuia," *Bar-Ilan* 12 (1972), p. 108.

69 R. Hananel's commentary on Baba Batra *ad loc.* in *Teshuvot Hageonim*, ed. S. Assaf (Jerusalem: 1942), p. 207. I would like to thank my friend Elimelech Westreich for this reference; also see Dinari, *op. cit.*, (note 68), p. 111.

pertaining to open spaces: "then the erection of walls could be made compulsory even in open spaces." However, the law pertaining to open spaces is only mentioned incidentally, and is certainly not "redacted Talmud," since Maimonides' interpretation and conclusion are not even mentioned, and it would not be difficult to come up with other logical accounts of the distinction between the law pertaining to open spaces and that pertaining to gardens.[70]

In any event, although Maimonides could have justified himself here, as he did when queried about the punishment for forbidden intercourse, by arguing that his ruling was in accordance with the redacted Talmud rather than the talmudic debate, he instead chose to offer a logical explanation for it. We can conclude that this response reveals his basic approach to resolving conflicts between incompatible sources. Within this framework, at least, in those instances where Maimonides deviates from the standard methods of resolving such conflicts, logical reasoning is a key factor.[71]

---

70 See *Tosafot* on Baba Batra *ad loc.* s.v. *gina shani.*

71 It is not difficult to find additional illustrations of the centrality of legal reasoning for Maimonides. Let us examine two such examples.

1) On the question of whether a witness can serve as a judge, Maimonides rules in accordance with the opinion of R. Tarfon that only a witness who actually testifies is barred from being a judge. He thus rejects the opinion of R. Akiba that even a potential witness, namely, one who saw the act but does not testify, is disqualified (*Code*, Laws Concerning Evidence 5:5). This conflicts with the principle in the Talmud (bEruvin 47b) that in a controversy between R. Akiba and one of his colleagues, the law is according to the opinion of R. Akiba; it is also in direct opposition to the passage in the Talmud which discusses the question (bRosh Hashana 25b). Cf. Benedict, who quotes additional sources, including the Jerusalem Talmud, which he regards as redacted Talmud, unlike the discussion in bRosh Hashana. He argues that Maimonides based his decision upholding the view of R. Tarfon as opposed to that of R. Akiba on these sources (*op. cit.* [see note 4], pp. 69ff.).

However, it is hard to believe that this was Maimonides' only reason for ruling at variance with the accepted principle for decision-making, and with bRosh Hashana, which cannot, from a logical point of view, be characterized as "incidental remarks" or "talmudic debate," as defined above. It is more likely that Maimonides is inclined to adopt the opinion of R. Tarfon on the strength of simple logic – the witness is barred from judging only if he is actually testifying. He thus prefers the words of the Mishnah as interpreted by the Jerusalem Talmud to the conclusion of the discussion in the Babylonian Talmud.

4. Maimonides' Attitude to Biblical Exegesis

Another area in which the role of logic in Maimonides' decision making is pronounced is that of the evaluation of laws derived by the Sages from the Bible by standard exegetical methods.[72]

i. Biblical and Rabbinic Exegesis

A survey of those laws in the *Code* falling into the category of exegetically-derived laws reveals that some are regarded by Maimonides as "biblical," others, as "Rabbinic." What criterion does he use to make the classification? The answer is clear: if, on the basis of logic and common sense, a teaching seems to follow from the plain meaning of the biblical text, or failing that, if it fits the plain meaning of the text, it is classified as biblical. If the teaching appears remote from this simple meaning or cannot be construed as clearly encompassed in it, the teaching is considered to be an allusion (*asmakhta*) only, and is regarded by Maimonides as Rabbinic rather than biblical.

---

2) Maimonides differentiates between the Temple and Jerusalem, which were "sanctified for that time and for eternity," and the rest of the land of Israel, which "was not sanctified for eternity" (*Code*, Laws Concerning the Temple 6:16). This is contrary to the conclusion of the Talmud (bShevuot 14a; Megila 10a) which does not uphold the distinction. R. Elijah of Vilna considers the matter and appears to conclude that the talmudic passages in question are not "redacted Talmud" (*Commentary on the Shulhan Arukh*, Yore Deia 331:6). Benedict characteristically rejects R. Elijah's position. Quoting the Tosefta, which upholds a view contrary to that of the Gemara, he contends that it constitutes redacted Talmud (*op. cit.*, pp. 137ff.).

Although this is conceivable, I believe that we should give precedence to what Maimonides himself tells us about the matter:

> Why do I say that the Temple and Jerusalem were sanctified for that time and for eternity, while the rest of the land of Israel ... was not sanctified for eternity? Because the former were sanctified by the divine presence, which does not lapse ... but the sanctity of the land of Israel with regard to sabbatical dues and tithes has to do with its having been conquered.

I would argue that these remarks, in which Maimonides uses reasoned argument to justify the distinction between the two types of sanctity, may well point to the factor which induced him to rule as he did. (Cf. Ravad *ad loc.*: "This is his personal theory; I do not know where he got it.")

72 See note 3; and Loewinger, *op. cit.* (note 4), pp. 34ff.

ii. Maimonides' Account in the *Book of Precepts*

We can learn more about this criterion from Maimonides' remarks in the *Book of Precepts* (*Sefer Hamitzvot*). In setting out the second methodological principle [in his terminology, the second Root (*shoresh*)], in which he states his well-known rule that a law derived using one of the "thirteen exegetical modes" or other rules for biblical exegesis, is not biblical, he also addresses the question of the criterion under discussion:

> Their lack of knowledge has already had even graver results, namely, when they come up with an exegesis of some biblical verse which makes it an obligation to perform or refrain from certain acts, which are indisputably Rabbinic injunctions, they count them among the precepts, though the **plain meaning of the biblical verse** does not command any of them, and in spite of the rule we were taught by the Sages, who said that the biblical verse must not be understood in a manner overstepping its plain meaning.[73]

Similarly, in the eighth Root, Maimonides distinguishes between a scriptural "nay" in the sense of an injunction or warning, and that of simple negation. He brings the following example:

> It has also been said that "fare not as Korah, and as his company"[74] is a negation, and the Sages explained it as a negation. ... in spite of this fact it is put differently in the Gemara in Tractate Sanhedrin: Those who sustain dissent transgress a negative precept, as it says in the Bible: "fare not as Korah, and as his company"[75]; [however,] this is merely an admonishment; the plain meaning of the biblical verse does not warrant the conclusion that it is an injunction.[76, 77]

iii. Terminology Related to Exegesis

Examination reveals that Maimonides' attitude to the types of exegesis distinguished above is also reflected in the *Code*, specifically, in the terminology used to articulate the connections between the laws and the Bible.[78]

---

73 Maimonides, *Sefer Hamitzvot*, trans. J. Kapah (Jerusalem: 1971); in English, *The Commandments*, trans. C.B. Chavel (New York: 1967).

74 Num. 17:5.

75 bSanhedrin 110a.

76 *Sefer Hamitzvot*, *op. cit.* (note 73), p. 29.

77 For more on this, see below.

78 See D. Rappel, "Darkei Hahesber Hadidacti Bemishne Tora Larambam," in *Mikhtam Ledavid, Sefer Zikaron Lerav David Ochs* (Ramat Gan: 1978), p. 291.

When a law follows from the plain meaning of the biblical text, or could be construed interpretatively as corresponding to this plain sense of the text, Maimonides quotes the scriptural verse and **adds an explanation**. For example,

> It is a positive biblical commandment that an alarm be sounded with trumpets whenever trouble befalls the community. For when Scripture says: Against the adversary the oppresseth you, then ye shall sound an alarm with the trumpets [Num. 10:9], this means, against anything that is oppressing you, whether famine, pestilence, locusts or the like, cry out and sound an alarm.[79]

Similarly:

> It is a positive biblical commandment to sanctify the Sabbath day through words, as it is said, Remember the Sabbath day to sanctify it [Exod. 20:8], meaning, remember it by praise and sanctification.[80]

But if the exegesis is further from the meaning of the biblical text, and cannot be inferred from it by the usual modes of interpretation, then Maimonides uses special formulas, such as "they have learned from tradition," etc. For example:

> By biblical law, relatives are ineligible to act as witnesses, as it is said: Fathers shall not be put to death for their children [Deut. 24:16]. **They have learned from tradition** that this negative command encompasses both the law that a father cannot be put to death by the evidence of his son, and, that a son cannot be put to death by the testimony of his father.[81]

And similarly:

> If a court is divided, some voting for acquittal and others for conviction, the majority opinion is followed. This is a positive biblical command … . What does this refer to? Monetary law … . But regarding capital cases, if, with respect to the same offender. … the majority is for conviction, he is not put to death unless those in favor of conviction exceed those in favor of acquittal by two. **They have learned from tradition** that this is what the Bible warned against by the injunction, "Thou shalt not follow a multitude to do evil" [Exod. 23:2].[82]

79 *Code*, Laws Concerning Fast Days 1:1.
80 *Code*, Laws Concerning the Sabbath 29:1.
81 *Code*, Laws Concerning Evidence 13:1.
82 *Code*, Laws Concerning the Sanhedrin 8:1.

The following question immediately arises: why does Maimonides consider these to be biblical laws, since they do not follow directly from the plain meaning of the verses cited? The answer is found in the second Root of the *Book of Precepts*, where Maimonides explains that if the Sages clearly state[83] that a law arrived at by exegesis is a biblical law, then it must be regarded as such, even if the interpretation cannot be inferred from the plain meaning of the biblical text. The term we have been looking at, "they have learned from tradition," and a similar expression, "it has been handed down," are used in the *Code* to refer to both biblical and Rabbinic commands.[84]

iv. The Precepts of Prayer and Emulation of Divine Conduct

In light of the above, it is particularly instructive to examine Maimonides' views on two unique commandments: the injunction to emulate divine conduct, and the injunction to pray.

Maimonides discusses the obligation to emulate divine conduct at length:

> We are commanded to follow these moderate ways, which are good and upright, as it is written, "you shall walk in His ways" [Deut. 28:9]. The meaning of this is understood thus: As He is called gracious, you should also be gracious, as He is called merciful, you should also be merciful.[85, 86]

---

83 This also appears to be true if it is clearly implied, though not stated.

84 In contrast to the examples we looked at, where the formula "from tradition" refers to biblical laws, at Laws Concerning Repose on the Tenth of Tishri 1:5 and Laws Concerning Forbidden Foods 17:5, it is used to refer to Rabbinic injunctions; similarly, the term "it has been handed down" is used to refer to a biblical law in the example we looked at from Laws Concerning the Sanhedrin, but in Laws Concerning Corpse Uncleanness 5:5, it denotes a Rabbinic command.

85 *Code*, Laws Concerning Moral Dispositions and Ethical Conduct 1:1-6.

86 Cf. *Sefer Hamitzvot*, Positive Command Eight: "The eighth precept we were commanded to obey is to emulate Him to the best of our ability, as He said, 'you shall walk in His ways,' **and this is explained**, 'as He is called merciful.'" As Maimonides states explicitly, his sources are Sifre Deut., Ekev 29 (p. 114 in Finkelstein ed.), and bShabbat 133b. Maimonides' position on actions such as visiting the sick, comforting the bereaved, and other deeds that reflect divine attributes, is complex. He seems to assert here that all such actions fall under "you shall walk in His ways," and should therefore be regarded as biblical injunctions. (However, in bSota 14a, the Sages interpret "you shall follow the Lord your God" the same way – as commanding emulation of the Lord's attributes, citing visiting the sick as an example. Maimonides himself hints at this as a source in the passage in question, that is, in positive command Eight of the *Book of Precepts*.)

Elsewhere, Maimonides states:

> It is a positive commandment to pray daily, as it is written, And you shall serve your God [Exod. 23:25] – they have learned from tradition that here "service" means prayer, as it is written, And serve Him with all your heart [Deut. 11:13]. The Sages said, What is the service of the heart? It is prayer.[87]

Scrutiny reveals that in both of these laws, Maimonides uses expressions which indicate the **distance** of the interpretation from the plain meaning of the biblical verse. In the first law, he states, "The meaning of this is understood thus"; in the second, "they have learned from tradition." Nevertheless, he defines both as biblical commands, even though there is no indication in the Talmud that the Sages regarded them as such.[88]

But in the *Code*, Laws Concerning Mourning 14:1, these same acts are described as Rabbinic injunctions! (Cf. Loewinger, *op. cit.*, p. 42, who finds this perplexing.)

Apparently, these actions should be understood from two different perspectives. To be sure, one who performs an act of loving-kindness is imitating divine conduct, and thereby fulfilling the biblical command, "you shall walk in His ways." But the particular actions of visiting the sick, comforting the bereaved, and so on, in and of themselves, are mandated by Rabbinic decree. Maimonides makes this very point in the second Root (see text at note 73). Against those who construe such acts as mandated by biblical commands, he maintains:

> They think that every separate action is a distinct precept, and do not know that all these acts, and others which are like them are included in one explicit biblical command, Love thy neighbor as thyself [Lev. 19:18].

87 *Code,* Laws Concerning Prayer 1:1.

88 With regard to prayer, the opposite seems to be the case: it is implied by the Sages that prayer is not a biblical command. Indeed, most of the Early Authorities disagree with Maimonides on this point. Cf. *Kesef Mishne ad loc.* It should also be noted that it cannot be inferred from the Talmud (bBerakhot 20a–b) that prayer is a biblical command. The Mishnah says only:

> Women, slaves and minors are exempt from reading the "Shma" prayer ... but are obligated to observe the precepts of prayer, fixing a scroll (*mezuza*) to their doorposts ... .

There are a number of variant readings in the Gemara, but from Maimonides' commentary to the Mishnah (Kidushin 1:7) it can be ascertained that he did not

In my opinion, these precepts reflect singularly well the essence of Maimonides' approach, according to which it is appropriate, and perhaps even necessary, that these basic precepts,[89] which constitute the very foundation of the service of the Lord, be regarded as biblical commands. Hence, in spite of the fact that the exegeses used to establish them overstep the plain meaning of the biblical verses, and in spite of the fact that the Sages neither explicitly nor implicitly referred to these precepts as biblical injunctions, Maimonides nonetheless identifies them as such on the basis of his own cognizance and principles.[90]

v. The Source of the Injunction Against Perpetuating Dissent

One of the laws mentioned above – the injunction against perpetuating dissent – provides an interesting illustration of Maimonides' sensitivity to different modes of exegesis.[91] In the eighth Root, Maimonides quotes an adage of the Sages: "Those who sustain dissent transgress a negative precept, as Scripture says: "fare not as Korah, and his company,"[92] explaining that this is a mere *asmakhta*, an "allusion," because it does not correspond to the plain meaning of the text. Maimonides adds: "However, the warning against this

---

have the version which asserts "should you say that since, regarding prayer, it is said in the Bible, evening, morning and noon, it should be considered a positive temporally-governed commandment – therefore it is written to refute this," since he explicitly states there that prayer is a temporally-governed precept (women being liable only by dint of a Rabbinic decree to that effect). On the basis of the Mishnah alone it cannot be inferred that women are obligated to pray because prayer is a non-temporally-governed positive precept; cf. *Sefer Hamitzvot*, Heller ed., First Root, note 4.

89 Though these two commandments may not be the only two such precepts; see note 97 below.

90 On the unique importance of emulating His ways, see *Code*, Laws Concerning Moral Dispositions and Ethical Conduct, at the end of the first chapter; on the precept of prayer, he elsewhere states:

> Let me return to my point and state ... the sacrificial service of the Lord is a secondary goal; the other kind of worship – cries of prayer and supplication. ... is closer to the primary goal, and essential in striving to attain it (*Guide for the Perplexed* 3:32).

91 See the text above at note 76.

92 bSanhedrin 110b.

matter is included in another negative commandment, as will be explained at the appropriate place." In the *Book of Precepts*, he indeed cites the prohibition: "We are warned against lacerating ourselves as the idolaters do, as the Almighty said, Ye shall not cut yourselves" [Deut. 14:1].[93] He then adds:

> And the Sages have said that this negative commandment includes a warning not to form contentious sects or engage in strife, and said, "'Ye shall not cut yourselves' – do not form factions among yourselves."[94] The basic meaning of the verse is, however, as was explained when it was stated that this was an injunction against infliction of self-injury in mourning. The admonition against forming sects is a **form of interpretation**, as is the adage, Those who sustain dissent transgress a negative precept, as Scripture says: "Fare not as Korah, and his company" – this is also an interpretation. ...but the basic meaning of the verse is an admonition, and therefore it is a negative statement, but not a prohibition.

Thus Maimonides explicitly connects the interpretations of "ye shall not cut yourselves" and "fare not as Korah, and his company," and defines both as interpretations which differ from the plain meaning of the text. Nevertheless, in the *Code* he rules that the injunction against forming factions is a biblical prohibition, and adds the following comment: "This includes the prohibition against two courts of law in the same town, each with its own way of doing things, because this causes great dissension, as it is written: Ye shall not cut yourselves – do not form factions among yourselves."[95]

Presumably, Maimonides ultimately makes a pragmatic distinction between the two interpretations we have discussed here. While he regards the exegesis of the verse about Korah as a mere *asmakhta*, he regards that of the verse "ye shall not cut yourselves" as implied by the text[96] and therefore construes it as a biblical command.[97]

93 Negative commandment 45.

94 bJebamot 13b.

95 Laws Concerning Idolatry 12:14.

96 Or so it appears from his laconic explanation in the *Code*: "As it is written, 'Ye shall not cut yourselves' – do not form factions among yourselves."

97 The Talmud (bJebamot 13b) treats the injunction against forming factions at length. Its focus is a discussion of the relationship between the House of Shammai/House of Hillel controversy over exemptions from levirate marriage, and the injunction against factions.

The Talmud also cites a dispute between Abbaye and Rava: Abbaye argues that "the prohibition against two courts of law in the same town refers to two courts,

vi. The Disqualification of Women from Serving as Witnesses

The case of another exegesis put forward by Maimonides as a biblical law is very different, however. I refer to his law disqualifying women from serving as witnesses. Maimonides writes:

> Women are disqualified from serving as witnesses by biblical law, as it is written: "At the mouth of two witnesses" [Deut. 17:6] – in the masculine gender but not the feminine.[98]

The *Kesef Mishne* commentary notes that the Talmud derives the disqualification of women by "verbal analogy," one of the thirteen traditional modes of exegesis,[99] while the argument from Maimonides' source, the "two witnesses" verse, "is weak, since the entire Bible is written in the masculine gender."

In my opinion, there is no doubt that although the Gemara derived this law using one of the prescribed modes of interpretation, and although according to Maimonides' criteria it is definitely not a biblical law, nevertheless, it is evident from the Mishnah and the discussion in the Gemara that the law is presented as a biblical law in spite of its exegetical origins.[100] Accordingly, as

---

one of which rules in line with the House of Shammai, and the other, in line with the House of Hillel"; while according to Rava, "it refers to a single court in a town, within which one group rules in line with the House of Shammai, and the other, in line with the House of Hillel." It would be wrong to assert that because of this passage, Maimonides felt bound to classify "Do not form sects" as a biblical command even though it does not follow from the plain meaning of the biblical verse, since this conclusion does not necessarily follow from the pericope. It is more likely that here too, as in the cases of the precepts of emulation of divine conduct and prayer, he decided according to his personal outlook as a result of his sense of the importance of the injunction. In one of his responsa he makes some very strong statements on this matter (*Responsa*, Blau ed., part 2, #329, p. 597).

It is also interesting that he rules in accordance with Abbaye's opinion here rather than Rava's, contrary to the standard talmudic decision-making rule, and in the face of what seems to follow from the talmudic discussion. On this, see the *Kesef Mishne* and *Mishne Lamelekh* commentaries on the *Code ad loc.* In my opinion, this has to do with the role played by logic in Maimonides' decision-making; however, the matter cannot be pursued further here.

98 *Code*, Laws Concerning Evidence 9:2.

99 bShevuot 30a. A similar, though not identical, account is found in Sifre Deut. on Judges, Finkelstein ed., par. 190, p. 230.

100 See the Mishnah *ad loc.*: "It is customary that men and not women take the oath of witness."According to Rashi, "it is written "and he gives witness" – what is

Maimonides rules in the second Root of the *Book of Precepts*, this law has to be accepted as a biblical injunction. But since the Talmud does not present this law as following **directly** from a biblical verse, but rather, as the indirect conclusion of a complex "verbal analogy," Maimonides prefers to cite his own source for the law, namely, the verse "at the mouth of two witnesses,"[101] though the argument is rather strained.

5. Conclusion

The significance of the various points that have been argued comes together when the article is considered as a whole, revealing the connections between the different subjects covered. This review clarifies both the particular cases considered and the main theme: the importance of logic in Maimonides' decision-making.

First we noted the role of logic in the rationales adduced for different laws in the *Code*. To this end, we presented an explicit explanation provided by Maimonides himself: his declared intention "to bring the laws closer to reason, or to what is most common," even if this necessitates deviation from the Talmud. We have tried to show that Maimonides carries out this program even in cases where the proffered rationale has actual legal consequences, which indicates, I believe, that Maimonides did not intend his rationales to be construed as law. However, here our distinction between internal-legal and external rationales comes into play: while the latter type of rationale – external objectives the law seeks to achieve – should not influence the application of the law, internal rationales, by their very nature, are integral parts of the laws they explain, and hence, indeed govern their application.

---

stated regarding who is qualified to serve as a witness is that women are disqualified from serving as witnesses, as is deduced in our chapter" (Rashi *ad loc.* s.v. *shvuat haeidut*). The Gemara *ad loc.* states: "What is the source of this custom? The Sages taught, And the two men stood."

101 Loewinger suggests that Maimonides regards verbal analogy as "a mere indication" that the scriptural term "two witnesses" applies only to men (*op. cit.*, p. 42). The Gemara and Sifre do not support this suggestion. Were Loewinger's argument valid, it would mean that Maimonides' formulation is imprecise and the main argument missing. Note that the Radbaz, in his commentary *ad loc.*, asserts that Maimonides apparently upheld the verbal analogy in Sifre [see note 99]; the Radbaz himself seems to have found it somewhat problematic, and hence added, "But this requires further study."

In the second section of the paper we discussed a more complex and significant subject: the rules for deciding between conflicting sources or opposed opinions. We attempted to demonstrate that here too logic plays a major role; indeed, in the process of enumerating the different laws, Maimonides makes extensive use of logic in deciding between conflicting legal sources. This conclusion was reached inductively: there are a considerable number of laws in the *Code* that are difficult to account for; Maimonides' principle endorsing "redacted Talmud" over "talmudic debate," articulated in a responsum, does not help in clarifying most of these problematic laws. However, in almost all cases, it becomes apparent upon reflection that simple logic supports Maimonides' rulings. (In this regard we also cited a responsum by his son, R. Abraham, who explained his father's decision in a dispute over the punishment for certain types of forbidden intercourse, in this spirit.)

The centrality of logic to Maimonides' teachings is most pronounced in the third area we examined – Maimonides' stance on laws established by exegesis. We saw that he systematically distinguishes between exegetical laws based directly on, or compatible with, the plain meaning of the biblical verse in question, and those requiring more complex exegesis. He considers the former tantamount to biblical laws, the latter, Rabbinic. Given that this critical distinction is made solely on the strength of logical considerations, it seems fair to extrapolate and conclude that logic is used by Maimonides in other spheres as well.

This is not to deny that it is not always easy to decide whether a given interpretation indeed follows directly from the verse's plain meaning or instead rests on subtle exegesis. It is sometimes possible to detect some hesitation in this regard, even in Maimonides. We noted, however, that in addition to logic, Maimonides' faith and religious outlook also play a decisive role in shaping his understanding of various precepts, at times directing him to regard a law as biblical, even if the interpretation of the text on which it is based deviates somewhat from its apparent meaning.

In conclusion, it appears that Maimonides' declared intention of "bringing the laws closer to reason," expressed with respect to his articulation of the rationales underlying various laws, also motivates other parts of the *Code*. The following remark made by his fiercest critic, R. Abraham b. David (Ravad), could with justice be applied to Maimonides himself: "Matters were clarified for us by the Tosefta and the Jerusalem Talmud ... but principally by true reason, which knows and discerns" (*Tamim Deiim*, 227).

6. Appendix: More on the Rationales for Laws in the *Code*

An important recent paper on the subject insightfully explores a number of instances in which Maimonides diverges from the reasons offered by the Talmud (R.N.A. Rabinowitz, *op. cit.* [note 4 above], p. 63). The author suggests that in such cases Maimonides assumes his reader is familiar with the rationale in the Talmud, and offers his own rationale to resolve some difficulty arising from it. Because a number of cases, including some of the examples discussed above, do not fit this model, I find this proposal problematic. Let me comment briefly on two of the laws Rabinowitz discusses, before considering the broader issue involved.

1. Laws Concerning the Daily Offerings and the Additional Offerings 1:3

> The daily afternoon offering can be slaughtered ... from six and a half hours into the day until the end of the day ... . Why is it delayed for two hours after the beginning of the time it can be slaughtered? Because of offerings tendered by individuals or the community. For it is forbidden to sacrifice any communal offerings before the daily morning offering or to slaughter any offering after the daily afternoon offering, with the sole exception of the Passover offering, **since it is impossible for all of Israel to bring their Passover offerings in the space of two hours.**

And: "The Passover offering is slaughtered only after the daily afternoon offering" (*ibid.*, 1:4).

Commentators on Maimonides have expressed puzzlement over the rationale he cites regarding the Passover offering, since in the Talmud (bPesahim 58b) an exegesis is brought to explain the fact that the daily offering precedes the Passover offering: "as it is stated, the sacrifice [the Passover offering] is in the evening and afternoon; it is delayed until after the sacrifice regarding which it is not stated, in the evening, but only, in the afternoon [the daily offering]." The *Lehem Mishne* commentary *ad loc.* suggests that were Maimonides' pragmatic rationale valid, it would make more sense to allow the Passover sacrifice to be slaughtered both before and after the daily sacrifice. Furthermore, on the line of reasoning advanced by Maimonides, would it not be more practical to defer the afternoon sacrifice in order to enable the Passover sacrifice to precede it? See Rabinowitz, *op. cit.*, pp. 166–67; and cf. Loewinger, who, pointing to these difficulties, says that Maimonides' rationale was offered solely to make the law more palatable (*op. cit.*, pp. 29–30).

Rabinowitz's theory is that Maimonides intended this rationale as an answer to a difficulty arising from the rationale in the Gemara: why is the daily sacrifice postponed by two hours, and no longer? Maimonides explains that it cannot be postponed any longer, because the Passover sacrifice takes more than two hours.

This interpretation seems forced. From the context of the law as cited it is obvious that Maimonides is replying only to the question of why the Passover offering is not sacrificed **before** the daily offering, as are all other sacrifices. His answer is that it is impossible to sacrifice the offerings of everyone in the community within the space of two hours. This raises another question: if so, why not sacrifice the Passover offering both before and after the daily offering? Maimonides answers this question in the next law: "The Passover offering is slaughtered only after the daily afternoon offering." That is, since the Passover offering cannot be completed before the daily offering, the biblical stipulation is that it should be sacrificed only after the daily offering. Maimonides did not consider the possibility of postponing the daily offering for more than two hours. Perhaps he believed it unseemly to delay the offering too long. In any event, as is evident from the laws that follow, sacrifice of the Passover offering indeed took more than three hours.

This example calls attention to a highly important distinction in the *Code*, of relevance not just here, but with respect to many other laws as well: the distinction between a law's source – Maimonides frequently brings an exact biblical source for the laws he enumerates (as a rule, the source is indicated by such phrases as "as it is stated") – and its rationale: the logical reason which explains why Scripture fixed the law in a particular form.

A good illustration of the distinction is provided by the following law, concerning the injunction against selling a house in the land of Israel to a Gentile, for which Maimonides notes both a source and a rationale. He writes: "And why is it not permitted to sell to them? As it is stated, Do not favor them – do not let them have a foothold [Deut. 7:2]. If they do not gain possession of land, their sojourn is temporary" (*Code*, Laws Concerning Idolatry 10:4). (Another good example, the injunction against praising them, follows.)

Maimonides does not do so in every case, however, and most often notes only the source: see, for example, Laws Concerning the Study of the Torah 1:1 and Laws Concerning Idolatry 10:1; also see Laws Concerning the Sanhedrin 20:1, on the injunction against punishing on the basis of circumstantial evidence, where Maimonides brings the source only: "and about such things it is stated, 'and the innocent and righteous slay thou not' [Exod. 23:7]." On the other hand, consider the example found in *The Book of Precepts*, negative precept 290, where Maimonides explains the underlying rationale for this same injunction at length. And see note 9 above.

Another example of a law for which no source is cited is the prohibition against condemning a frontier town to destruction, discussed above (Laws Concerning Idolatry 4:4). In this case, too, Maimonides brings the rationale underlying the prohibition – to prevent an enemy invasion of the land of Israel – but not the biblical source. Note, however, that Maimonides is not trying to claim that this rationale is itself the source of the law, he simply does not consider it appropriate to indicate the source in question.

Did Maimonides assume that readers would be familiar with the scriptural sources cited in the Talmud, as Rabinowitz suggests? It is not possible to be sure, but from remarks in his introduction to the *Code*, this seems unlikely. This was perceived by R. Menahem Hameiri: "This work organizes the wisdom of the Talmud. ... it almost seems to have been intended to make the Talmud unnecessary. ... as the Master of blessed memory reveals in his introduction" [*Beit Habehira* on Berakhot, Dikman ed. (Jerusalem: 1965), p. 25]. However, the matter is not germane to our present argument.

2. Laws Concerning Slaves 1:6

The next example has bearing on both the question of Maimonides' rationales for laws and his treatment of sources; see note 1 above.

> It is forbidden to work any Hebrew slave strenuously. What constitutes strenuous work? Work which has no fixed limits and work which is unnecessary .... Hence the Sages said that a master should not tell his slave, "Hoe beneath the vines until I come back," because he has not limited the time .... Nor should one say to him, "Dig up this place," where he is not in need of it, or command him even to warm or cool a cup of water for him if he is not in need of it.

Commentators have asserted that the source of this law is Sifra on Lev., Behar 25:43:

> Thou shalt not rule over him with rigor. Do not say, Warm this cup for me, if it is not needed; Hoe beneath the vines until I come back. Lest you say, I do it for my needs – This is a matter for his conscience, as it is written, thou ... shalt fear thy God.

Maimonides makes a distinction between needless work and work lacking a time limit. This distinction is harshly criticized by R. Abraham b. David:

> This is his personal speculation, but the Sages do not mention the time limit as a reason, and said only that the master should not say "until I come back," since he might tarry and cause suffering, and this is what would weigh on his conscience… . But this has nothing to do with time limits. Furthermore, the slave will be nothing but a laborer… . But he is an eccentric author who says eccentric things.

According to Rabinowitz, Maimonides does not define two distinct types of strenuous work – needless work and work without a time limit – since "any work which has no time limit is not needed at present" (*op. cit.*, p. 171). He argues that Maimonides mentions the time limit because the *baraita* in Sifra referring to conscience, hinted at by the Ravad, seems to apply to both orders: the command that the slave warm a cup unnecessarily, and the command that he work until the master returns.

However, it is difficult to see why the latter is a matter of conscience, but for the time limit element. Clearly, though, the text of the *Code* does not support this explanation, for it explicitly states, "What constitutes strenuous work? Work which has no fixed limits and work which is unnecessary." This implies two distinct kinds of work. The distinction is maintained at the end of the law: "The slave does for him only work with fixed limits for which there is a need."

Therefore we should simply accept the fact that Maimonides and the Ravad disagree on the interpretation of the *baraita* in Sifra. According to the Ravad, the appeal to conscience applies to both orders, but according to Maimonides, it applies only to the command that the drink be warmed. What is common to both interpretations, however, is the effort to overcome the textual difficulty in Sifra, where "hoe … until I come back" seems to be incompatible with what follows, as explained above. Cf. *Lehem Mishne ad loc.* As Rabinowitz notes, Maimonides quotes Sifra in the *Book of Precepts*, Negative Precept 259, but omits the phrase "hoe … until I come back." I believe it probable that at the time he wrote this, Maimonides was using a version of Sifra in which the phrase was missing, and only later did he see the full version, at which point he explained it as he did in order to resolve the textual difficulty.

It is also possible that "hoe … until I come back," the second phrase in Sifra, is a **later addition**, because, as we said, Maimonides himself omits it in the *Book of Precepts,* and it is found in neither *Sefer Yereim Hashalem* nor

Sefer Mitzvot Gadol. In the former, the text reads "and it was taught, Thou shalt not rule over him with rigor – he should not say to him, warm that cup, when it is not needed. And if you say, I do so for I am in need of it... ." (p. 199). The version in the *Semag*, negative commandment 174, is also cited in *Hagahot Maimoniot ad loc*. It is obvious that these authors were working with a version of the Sifra containing only the first phrase, and in any case, it is unlikely they omitted the second phrase to register disagreement with Maimonides, as R. Benedict (*op. cit*., p. 270) contends.

R. Benedict also suggests a different source for the law, arguing that Maimonides' "work which has no fixed limits" comes from R. Moses the Preacher, who, regarding the slavery of the Israelites in Egypt, quotes Moses as describing their strenuous toil as "work with no fixed limit." As we saw, this exegesis is unnecessary.

Also note that Rashi's commentary on Lev. 25:43 indicates that he does not appear to distinguish between the activities referred to by the two orders in question:

> Thou shalt not rule over him with rigor – unnecessary work in order to torment him. Do not say to him, warm that cup for me... hoe beneath the vines until I come back, lest you say, no one can tell whether it is needed or not, and I tell him it is needed... it is a matter for your conscience.

But as noted, Maimonides argues that there is most definitely a distinction. *Sefer Hahinukh*, which ordinarily follows Maimonides' decisions, indeed explicitly asserts that needless work and work without a time limit are two different things, see Precept 347, Chavel ed.

CHAPTER 6

EVASION OF THE LAW IN THE TALMUD

Shmuel Shilo

1. Introduction

Even a superficial perusal of the talmudic sources that mention evasion reveals two distinct attitudes – some evasions are prohibited, others are condoned. Instances of the latter include passages such as:

> One may evade [paying tithes] on his grain by concealing it in the chaff, so that his animal might eat it.[1]

and

> One can increase the amount of grain in the hay and so evade paying tithes.[2]

There are numerous statements in this spirit. However, in other cases, the ruling is "as long as he does not evade; if he evades, they are impure."[3] Is it possible to arrive at a reasonably clear picture of the Sages' attitude to evasion from the talmudic sources? Asheri, for example, is aware of the problem, and notes, with reference to evasions of Rabbinic injunctions, that

> All types of evasion are not equal, some are permitted to anyone, others are permitted only to scholars ... some are forbidden altogether and the Sages regard them as serious offenses if perpetrated deliberately ... on others, opinion is divided.[4]

Asheri points out the problem but offers no solution. In my opinion, however, the talmudic sources can be systematically analyzed and explained, and we need not adopt the Rashba's view that "one cannot compare different kinds of evasion."[5]

---

1 bBerakhot 31a, bPesahim 9a, bAvoda Zara 41b, bNida 15b.
2 jBerakhot 5:1 (8d).
3 Tosefta Toharot 10:5, 10:12; Tosefta Para 4:13; Tosefta Keilim, Baba Metzia, 5:14; Tosefta Ohalot 8:4.
4 *Piskei Harosh*, Shabbat 20:5.
5 *Hidushei Harashba*, Beitza 11b s.v. *R. Ada bar Ahava.*

Before commencing our investigation, an important distinction, which can help clarify the problem, should be made. Not infrequently, the term *haarama*[6] is used in the talmudic literature as a synonym for fraud,[7] a subject not treated here. Several seemingly contradictory sources can be reconciled by distinguishing between evasion and fraud.

There is no simple way to specify the precise meaning of "evasion" as it is used in the various tractates of the Talmud; indeed, its meaning can only be ascertained by thoroughly examining each passage in which it appears. While previous scholars have been aware of this, we should exercise caution with regard to their conclusions – for example, the view expressed by Maimonides in his *Commentary on the Mishnah*: "Permissible tricks are called evasion, forbidden tricks, cheating."[8] Complicating the subject further is the fact that the researcher must take into account numerous other cases which are not explicitly specified to be instances of evasion but which do, in fact, involve evasion.

All too often, talmudic scholars conflate two distinct concepts: evasion and legal fiction. While a common motivation may underlie use of evasion and legal fiction, namely, the desire to bring about an acceptable outcome, they are nevertheless quite different. This paper focuses on evasion of the law, and not on legal fiction. Atlas defines legal fiction as follows:

> Essentially, fiction is a change within the law to the effect that a particular phenomenon should be regarded as if it were something else; having been identified, they are to be treated as one … . The substance of legal fiction, both in Jewish law and in general, is completely different from evasion. In legal fiction there is no avoidance of a law by finding a loophole to circumvent it, but rather, extension of a law's scope to cover a greater area than originally intended. That is to say, the law is extended to apply to phenomena completely different from those to which it originally applied. A legal fiction exists when a given event is legally defined "as if" it were a different event, and judged accordingly. Thus, legal fiction does not call for a change in the course of events; rather, it is a legal creation

6 In the Bible, the root ARM sometimes denotes cleverness and wisdom without a negative connotation, e.g., "To give prudence (ARMa) to the simple, to the young man knowledge and discretion" (Prov. 1:4). But even in the Bible, its usual meaning is, as in contemporary usage, artfulness and cunning, with a negative connotation, e.g., "Now the serpent was more subtle (ARuM) than any beast of the field" (Gen. 3:1).

7 E.g., bShabbat 38a, jKidushin 3:4 (64a), jPeia 1:5 (16c).

8 Maimonides, *Commentary on the Mishnah*, Temura 5:1.

> instructing us to consider a given situation according to a law which was originally intended to cover a different situation.[9]

The crucial concept in the definition of legal fiction, then, is "as if" – a given situation is regarded as if it were another.[10] Evasion of the law, on the other hand, is a way of avoiding the consequences of applying a law to a given case by finding a loophole in the law itself, making the case, from the point of view of the law, fall into a different category, one in which legal outcome is different. My paper, as I said, deals only with evasion, or avoidance of the law, and not with legal fiction.

It could be argued that the law can never be evaded, since any action must be either permissible and legal, or forbidden and illegal, and there is no third possibility. What is conceivable, however, is an act that appears to comply with the letter of the law, but is clearly contrary to the intention of the legislator. In such a case it is likely that from the legal perspective the act must be defined as breaking the very law it purports to uphold. In Holmes's words,

> We do not speak of evasion, because when the law draws a line, a case is on one side of it or on the other, and if on the safe side is none the worse legally that a party has availed himself to the full of what the law permits. When an act is condemned as an evasion what is meant is that it is on the wrong side of the line indicated by the policy if not by the letter of the law.[11]

Generally, recourse to legal fiction makes it possible to extend the scope of a law to cover something not previously governed by that law. Evasion of the law, on the other hand, is usually undertaken to remove a particular situation from a law that would ordinarily apply. Note that fiction does not mandate change in the state of things, but is a legal device which states that a given situation will henceforth fall under the scope of a certain law, though that law was never intended to apply to it.

On examination, talmudic law reveals two main types of evasion, which can be further subdivided. The first type covers cases in which a well-known law is used to circumvent a law which would ordinarily apply. Under this

---

9 Atlas, S. *Netivim Bemishpat Haivri* (New York: 1978), pp. 224, 227–228.

10 On legal fiction, see L. Fuller, *Legal Fictions* (Stanford: 1967); and H. Vaininger, *The Philosophy of "As If" – a system of the theoretical, practical and religious fictions of mankind* (London: 1924), originally published in Germany in the nineteenth century. Fuller is critical of the translation (Fuller, *ibid.*, p. 44, note 3).

11 Buller v. State of Wisconsin 36 S. Ct. Reporter 473 (1916), p. 474.

category we find such subcategories of evasion as transfer of ownership in order to evade the law, and transfer of ownership to non-Jews for the same purpose. The second type of evasion covers cases in which the element determining the permissibility of an act is the agent's subjective intention, with the evader arguing that his intentions are such as to render the act permissible, while in fact they are different. It is difficult to define this kind of evasion, which is closer to deception and fiction (though not legal fiction) than to the type of circumvention described above.

It is sometimes argued that a particular evasion mentioned in the Talmud does not fall under either category and requires a different classification. However, I believe the analysis presented here can, in fact, account for all, or almost all, of the instances cited.

It should be noted at the outset that this discussion will not cover acts which originated as evasions but by the talmudic period had already become established legal institutions, by enactment or by other means. An example is the *prosbol*, which seems to have originated from what we have defined as evasion.[12] It could also be argued that agency in litigation is based on evasion.[13]

2. The First Type of Evasion: Invoking One Law to Circumvent Another

A. Tithes (*maaserot*) and Priestly Gifts (*terumot* and *hala*)

One area of Jewish law in which cases of evasion are relatively numerous is that dealing with tithes, and heave and dough offerings. Various examples that fall into the first category – circumvention – are found in the sources, as are what seem to be differing, and even contradictory, views on evasion, and a number of arguments for and against it. This area thus provides a superb illustration of the depth and complexity of the concept of evasion; indeed, the material pertaining to tithes and heave and dough offerings constitutes the best case study for exploring evasion in the Talmud.[14]

---

12 Much has been written on the *prosbol*; see N. Rakover, *Otzar Hamishpat* (Jerusalem: 1975), pp. 152–153, and M. Silberg, *Talmudic Law and the Modern State*, trans. B.Z. Bokser (New York: 1973), pp. 38–41.

13. See Silberg, *op. cit.*, pp. 32–34; Rakover, *Hashelihut Vehaharshaa Bamishpat Haivri* (Jerusalem: 1972).

14 The laws of dough offerings should be included under the laws of heave offerings, according to bMeila 15b, on the basis of Num. 15:20 – "of the first of your

Tithes and heave offerings are due only after the crop is harvested and brought into the farmer's house in the customary way. According to biblical law, if the farmer brings his crop home in an unusual way, he is exempt from the obligation to set aside tithes and heave offerings for the priests. It is not hard to discern an opportunity for evading this obligation by intentionally deviating from the usual routine. Our question, then, is the Sages' attitude to one who intentionally deviates from his usual routine in order to evade this obligation.

In the laws of tithes and heave offerings, as in other areas, obligation is based on the ownership relation. It is interesting to inquire into the position of the halakha if a change of ownership is effected in order to evade the obligation to bring priestly gifts. One such strategy is to abandon property and then reclaim it; another is to use a technicality to transfer ownership to someone else (Jew or non-Jew) in order to gain exemption, strategies suggested in discussing the obligation to give priestly gifts and second-tithes.

i. Changes in Methods of Harvesting

A law that was apparently unanimously accepted, and was referred to in the Talmud as "a decided law"[15] (though the meaning of this expression is still not fully understood[16]) seems to be attributed to the Tanna R. Oshea[17]

---

dough ye shall set apart a cake for a gift (*teruma*)." But it has become customary to treat dough offerings as a separate subject. In Maimonides' *Code*, for example, the laws of dough offerings are not included in the laws of heave offerings but appear separately at the end of the laws of first fruit offerings (*bikurim*). We examine the laws of the dough offering below and compare dough offerings, heave offerings and tithes with regard to evasion.

15 bBerakhot 31a. Cf. Y. Emmanuel, "Maarim Adam Al Tevuato" *Hamaayan* 19 (1979), p. 34, and note 2; S. Kook, *Iyunim Umehkarim* (Jerusalem: 1959), vol. 1, pp. 72–73.

16 See Atlas, *loc. cit.* (note 9). Atlas argues that Rashi's interpretation of "decided law" as one "which is not questioned and there is no need to dwell on it," is improbable. Nor is Atlas's attempt to explain this expression convincing. Kook argues that a "decided law" is only "decided" for its proponents, but his evidence for this is similarly unconvincing, see Kook, *op. cit.* (note 15), pp. 72–76. Cf. a similar passage in jBerakhot 5:1 (8d), which speaks of "fixed laws" rather than "decided laws" in the Babylonian Talmud.

17 I am inclined to accept Ginzberg's interpretation of this pericope, namely, that the person involved was the Tanna R. Hoshea the Elder, a young contemporary of R. Judah the Prince, the editor of the Mishnah, and not the Amora of the same name; see *Peirushim Vehidushim Bayerushalmi* (New York: 1941–1961),

(or Hoshea): "One may evade [paying tithes] on his grain by concealing it in the chaff, so that his animal might eat it."[18] Another law (perhaps a variant of the previous one[19]) states in the name of R. Hoshea, "One can increase the amount of grain in the hay and so evade paying tithes."[20] We will not go into the various problems that arise here, such as whether these rules apply to food eaten between meals or just to meals,[21] and whether the mixtures are permitted for human consumption, or only for animal consumption.[22]

Evasion is explicitly permitted if the crop is brought home before being properly levelled off, since its harvesting has not been completed in the usual way. Nevertheless, on any interpretation of these sources and the relationship between them, it is clear that one is permitted – or at least able[23] – to avoid

---

part 4, pp. 119–120. But there is no general agreement on this point. The author of *Dikdukei Soferim* on bBerakhot 3la, leaves this an open question. One of Ginzberg's strongest arguments is the usage "*tani* [a *baraita* taught by] R. Hoshea" (jBerakhot, *loc. cit.*). It is strengthened by another pericope on an evasion permitted by R. Hoshea which also uses "*tani*":

> The Mishnah speaks of large vessels, but with regard to small vessels the law may be evaded and they may be cleansed by immersion. It is taught by (*tani*) R. Hoshea: A man fills an impure vessel from the well and evades and immerses it for cleansing. It is also taught: If his bucket or utensil falls into the well, he may evade and immerse them for cleansing (jTerumot 2:1 [41c]).

Cf. "Immersion of Utensils," below.

18 bBerakhot 31a, bPesahim 9a, bAvoda Zara 41b, bMenahot 67b, bNida 15b.

19 There is some controversy as to whether these statements are parallel versions of the same law or two distinct laws. Ginzberg (*loc. cit.*) and Rattner, *Ahavat Tzion Veyerushalaim* (Vilna: 1901–1913), Berakhot, p. 120, argue that the Babylonian Talmud differs with the Jerusalem Talmud on this. Cf. *Gilionei Hashas* on the pericope.

20 jBerakhot 5:1 (8d).

21 See the commentaries of the Early and Later Authorities, e.g., Rashi, *Tosafot*, R. Judah Sirleon, and Hameiri on Berakhot *ad loc.*; *Tosafot* and R. David Bonfils on Pesahim 9a s.v. *kedai*; and R. Solomon b. Adret, *Responsa*, part 1, #361.

22 See commentary by the Early and Later Authorities on this question, for example, Rashi, *Tosafot*, R. Judah Sirleon and Hameiri on Berakhot *ad loc.; Tosafot*, R. David Bonfils and *Hidushei Haran* on Pesahim 9a; *Hidushei Haramban* on Avoda Zara 41b; *Tosafot* on Menahot 67b s.v. *kedai*; *Tosafot* and *Tosafot Harosh* on Nida 15b s.v. *behemato okhelet*; R. Solomon b. Adret, *Responsa*, and R. Shlomo Goren, *Hayerushalmi Hameforash* (Jerusalem: 1961) on Berakhot 5:1.

23 On whether evasion is permitted a priori, or accepted post factum, see below.

these obligations by intentional actions, undertaken expressly for this purpose, that constitute artificial changes in the harvesting process. No reservations about such actions seem to be expressed.[24] But another source implies that an evasion with a similar intent, though legal, meets with the Sages' disapproval.

ii. Changes in How Produce is Brought into the House

> And Rabba bar bar Hana said in the name of R. Johanan who said in the name of R. Judah bar Ilai: Come and see that the later generations are not like the earlier generations. The earlier generations brought their produce home through the front door in order to be obliged to give tithes. The later generations bring their produce home through roof-tops and courtyards[25] and backyards to be exempt from tithes. For R. Yanai said: The produce is not subject to tithes before it has reached the front of the house, as it is written "I have put away the hallowed things out of my house". R. Johanan says that even the yard renders tithes obligatory, for it is written: "They shall eat in your courtyards to their satisfaction."[26]

If there is any doubt as to whether the earlier or later generations are preferred by R. Judah b. R. Ilai, it is dispelled by a preceding passage that quotes the same masoretic chain[27] – Rabba bar bar Hana, who said in the name of R. Johanan, who said in the name of R. Judah bar Ilai:

> Come and see that the later generations are not like the earlier generations. The earlier generations made their studies their permanent occupation and their work

---

24 In the original version of this law, the tone is not negative. It is possible that this evasion was disapproved of only later, see bMenahot 67b; and see below, "Selling the Dough to a Non-Jew."

25 See *Dikdukei Soferim*, Berakhot 35b, on "courtyards." Apparently, it would be more correct to delete "courtyards," as in a parallel discussion at bGitin 81a.

26 bBerakhot 35b.

27 This is the printed version; one of these wordings may be more correct: "Rabba bar bar Hana said in the name of R. Johanan" or "Rabba bar bar Hana said in the name of R. Judah bar Ilai"; see *Dikdudei Soferim ad loc*. In my opinion, the correct version is the latter. The anecdote is apparently based on jMaaserot 3:1, 50c, where it appears with a slight difference:

> R. Ula b. Ishmael said in the name of R. Lezer: Rabbi and R. Jose b. Judah brought their bushels in from behind the rooftops. R. Judah b. R. Ilai saw them and told them, Look at the difference between yourselves and the earlier generation. R. Akiba used to buy three different kinds for a penny to

> incidental, yet they succeeded in both. The later generations made their work their permanent occupation and their studies incidental – but succeeded in neither.[28]

Two conclusions follow: as in the ruling of R. Oshea, it is the end state of the process, in this case, that of bringing the produce into the house in an unusual manner, that gives rise to the exemption from tithes. Here, however, the tone is negative, which is not the case when the rule of R. Hoshea is quoted.[29] Recall that the Jerusalem Talmud, as we noted above (see note 27), reports that R. Judah b. R. Ilai witnessed reputable scholars practicing evasion by using rooftops, and criticized them for it.[30]

Another kind of evasion that falls under the category of circumvention makes use of expropriation and transfer of ownership. This kind of evasion pertains only to priestly gifts and second-tithes. As we shall see, there appears to be a range of different attitudes to such evasion. We begin by considering an example explicitly permitting evasion, and then move on to instances in which the permissibility of evasion is problematic.

iii. Transferring Ownership

There is a biblical obligation to consume second-tithes in Jerusalem. However, one who does not wish to carry his produce to Jerusalem may redeem it for its market value, add a fifth of its worth, take the money to Jerusalem, and spend it on produce, which must be consumed there.[31] However, if the produce is redeemed by someone else and not by the owner, the extra fifth need not be added.[32] In other words, if the owner of the produce and the owner of the money are two different people, there is no obligation to add a fifth. As R. Johanan says: "Any tithes and their redemption money that do not belong to the same person are exempt from the fifth."[33] It is not surprising to find that evasion by fictitious transfer of ownership of

---

> make himself eligible to give tithes from each kind, while you bring in your bushels from behind the rooftops?

See note 64 below.

28 bBerakhot, *loc. cit.*

29 This will be discussed below, see "Selling the Dough to a Non-Jew".

30 jMaaserot, *loc. cit.*

31 Num. 27:31, Mishnah Maaser Sheini 4:3, Mishnah Baba Metzia 4:8.

32 Pesikta Zutrata, Behukotai (Num., *loc. cit.*); bBaba Kama 69b, and Rashi *ad loc.* s.v. *tana tznuim*.

33 jMaaser Sheini 4:3 (4:4 [55a]). See note 64 below.

produce or money to someone else is discussed in the Mishnah. But the attitude of the Tannaim to this type of evasion is somewhat more surprising.

We find the following passage in the Mishnah:

> Second-tithes are evaded. How? A man says to his adult son or daughter, or to his Hebrew slave or bondmaid, "Take this money and redeem for you this second-tithe." But one must not say so to his minor son or daughter or to his non-Jewish slave or bondmaid, because their hand is like his own.[34]

That is, money is given to someone else to redeem the second-tithe, thereby enabling the owner to avoid paying the surcharge. After ownership of the money has been transferred, it no longer belongs to the owner of the second-tithe.

In the next Mishnah, evasion in the other direction is illustrated – transfer of ownership of the second-tithe produce to avoid paying the extra fifth.

> He stands on the threshing floor and has no money on him; he can say to his friend, Take this produce as a gift [and then] this is redeemed with money I have at home.[35]

Apparently, at first the custom was to use the evasion technique described in the first Mishnah, but "since swindlers have multiplied," it was superseded by that described in the second Mishnah.[36]

---

34 Mishnah Maaser Sheini 4:4; bGitin 65a; cf. jMaaser Sheini 4:2: Does not the fifth of the first exceed the increment of the other? R. Abin said: This is different because he can exempt himself from paying the extra fifth by evasion.

35 Mishnah Maaser Sheini 4:5, bBaba Metzia 45b–46a.

36 Cf. Tosefta Maaser Sheini 4:3:

> The duty of the extra fifth on second-tithes is evaded. How? A man says to his adult son or daughter or to his Hebrew slave or bondmaid, Take for you this money and redeem with it this tithe. But he must not say: Redeem for me this tithe. R. Joshua b. Korha said: At first this was the custom, but since swindlers have multiplied, one says to his fellow: These fruits are yours as a gift, and he continues, they are redeemed with fruit which I have at home. But he must not say, They are redeemed with money I have in my pocket, before the other had a chance to take hold of the fruit, or rent the place on which the fruit was laid.

The reason for the change from transferring money to transferring fruit is explained as follows: "At first they used money, but sometimes they took the money and ran away with it, so it was enacted that produce be used" (jMaaser

An explicit justification of this evasion – though its interpretation is controversial[37] – is found in the Jerusalem Talmud.

> R. Abun said: R. Lezer and R. Jose b. R. Hanina disagreed. One said: Why do we evade it? Because the verse referring to it contains a blessing.[38] The other said: Why do we redeem it at the lowest rates? Because the verse referring to it contains a blessing.[39]

It is clear from all these passages that evasion by fictitious transfer of ownership is permitted. Apparently, since the technical legal requirements of a transfer are satisfied, the outcome of the transfer is not disqualified, and it may be the case that no ethical blemish is incurred.[40]

Sheini 4:3, [4:5 55a]). The Jerusalem Talmud goes on to relate that even using this system there were cases of absconding, and yet another method of evasion had to be introduced.

37 See commentaries on the Mishnah and Jerusalem Talmud *ad loc.*; cf. *Responsa Hatam Sofer*, Orah Haim #62.

38 The biblical verse is:

> And if the way be too long for thee, so that thou art not able to carry it, because the place is too far from thee, which the Lord thy God shall choose to set His name there, when the Lord thy God shall bless thee; then shalt thou turn it into money, and bind up the money in thy hand, and shalt go unto the place which the Lord thy God shall choose (Deut. 14:24–25).

39 jMaaser Sheini 4:3 (4:5 55a).

40 The Hatam Sofer (*loc. cit.*) argues that the evasion is condoned to encourage the flow of currency, as opposed to produce, to Jerusalem, because the blessing obtains only when money is brought to Jerusalem. Cf. a ruling by the great jurist Learned Hand:

> A transaction otherwise within an exception of the tax law, does not lose its immunity because it is actuated by the desire to evade taxation. Anyone may so arrange his affairs that his taxes shall be as low as possible; he is not bound to choose a pattern which will best pay the Treasury: there is not even a patriotic duty to increase one's taxes (Helvering v. Gregory 51 F. 2d. 224 at 225).

Cf. Marvin A. Chirelstein, "Learned Hand's Contribution to the Law of Tax Avoidance," 77 *Yale Law Journal* (1967–68), pp. 440–474. This and other examples make it clear that Jewish law does not adopt the doctrine of "piercing the veil" as do other legal systems to various extents; unlike them, it is not inclined to annul transfers to "straw men."

iv. Abandonment

On the surface, the Sages seem to have a different attitude to evasion by abandonment than that evinced with regard to the evasion of tithes by transfer of ownership. On closer examination, however, we will see that rather than contradicting their previous position, the Sages' discussion of abandonment actually reinforces it.

By law, abandoned produce is exempt from tithes and priestly gifts.[41] The easiest way for an owner to avoid his obligation to set aside tithes and gifts for the priests is thus to abandon and subsequently repossess the produce.

As the following *baraita* shows, the Sages were aware of this stratagem:

> One who abandons his vineyard and early the next morning gathers the grapes, is exempt from tithes but liable to leave gifts for the poor (separated grapes, small bunches, gleanings and the corners of the field).[42]

To prevent this facile form of evasion by abandonment and repossession, the Sages stipulated that for abandonment to be valid it had to be declared in the presence of at least three people – so that one could take possession, if he so wished, and the other two could bear witness to the fact.[43]

The Sages used other enactments to hinder evasion of tithes. According to biblical law, abandonment is valid immediately after being declared.[44] But the Sages issued an enactment to the effect that abandonment takes hold only after three days, "because of the swindlers" who abandoned their fields and immediately repossessed them, to exempt them from tithes.[45] The enactment states that if one who abandons his property repossesses it within three days, it is as if he never abandoned it, and he must give tithes and heave offerings.

This enactment and its results were interpreted by commentators on the Talmud[46] in various ways, which need not be considered here; all agree, however, that its aim was to prevent evasion of tithes by abandonment and repossession.

---

41 Mishnah: Hala 1:3, Peia 1:6, Sheviit 9:1, Terumot 1:4, Maaserot 1:1.

42 bBaba Kama 28a, 94a; bNedarim, 44b; bHulin 134b; bTemura 6a; bNida 5la.

43 bNedarim 45a.

44 *Ibid.*, 43a.

45 Tosefta Maaserot 3:11; bNedarim 43a–45a; cf. jPeia 6:1 (19b).

46 See the following commentaries on bNedarim 45a: *Tosafot*, Rosh, Meiri, Ritba; *Tosafot* on bShabbat 18b s.v. *demifkara*; *Tosafot* on Pesahim 4c s.v. *midoraita; Tosafot* on Baba Metzia 30b s.v. *afkarei*; Ramban on Pesahim (the beginning); *Nimukei Yosef* on Baba Metzia, the end of the second chapter.

On the surface, this attitude seems to contradict the more "liberal" approach to the evasion of tithes and priestly gifts noted above, for it is a clear case of the Sages deliberately taking steps to eliminate loopholes and prevent evasion. But a closer look reveals that the contradiction is more apparent than real, and that with regard to the validity of evasion by transfer or relinquishment of ownership, there is no controversy at all. The very need for an enactment to alter the rules of abandonment shows clearly that the Sages saw no possibility, within the existing law, of invalidating the fictitious transfer of ownership on account of insincerity, because by law, transfer and abandonment acquire validity by fulfilling objective criteria, and hence cannot be invalidated on the basis of intention. However, since, from a religious–legal point of view, the practice of blatant evasion was undesirable and potentially dangerous, the Sages enacted new legislation in an attempt to prevent it.

Thus it is clear that in spite of the Sages' uneasiness over this particular kind of evasion, they felt compelled to accept it if it satisfied the halakhic requirements. Given that they could not find fault with the actions, which did, indeed, conform to the law, the only response open to them was to change the law to make evasion more difficult.[47]

Furthermore, there is some support for the view that the enactment which rendered the fields of non-Jews subject to tithes and priestly gifts came into being to prevent evasion by "*baalei kisin*"[48] who transferred ownership of their fields to non-Jews to exempt them from tithes and priestly gifts.[49]

---

47 Cf. Marluyn's position:

> It cannot be stated too emphatically that there can be no such thing as evasion of law; the phrase itself is completely devoid of any meaning whatsoever; any specific act on the part of any given citizen at any given time is or is not an infraction of the law as it exists at that time, and it is the judge's duty as interpreter of the law, to state whether or not that specific act did or did not constitute an offence against the law as it existed at the time the alleged offence was committed. ... if a law is so badly framed that it fails to fulfil its intent then it must be re-drafted [*First Principles of Human Law* (Capetown: 1954), p. 137].

48 It is not clear just who these "*baalei kisin*" were and what they intended to evade. The Early Authorities suggest various interpretations. Cf. Rashi on bMenahot 67a s.v. *gezeira mishum baalei kisin*; Maimonides, *Code*, Laws Concerning Heave Offerings 4:15.

49 bMenahot 67a.

However, we will not discuss this enactment here since it is not certain that this was in fact its purpose. It is possible that the purpose of the enactment was to prevent dubious transactions such as fraudulent transfers of property to non-Jews,[50] or that there was some other reason for it.[51] If, indeed, the former motive is behind the enactment, it serves to further demonstrate the fact that the Sages did not claim the authority to invalidate acts of deliberate evasion, and had to resort to passing a new law making non-Jews liable for tithes and priestly gifts, contrary to biblical law.

v. Reducing the Amount of Dough

A further example of a valid act of evasion which the Sages regarded as improper conduct, and even explicitly prohibited – though it remained legally valid – is evasion of the priestly gift of the dough offering (*hala*). The duty to set aside a portion of one's dough as a priestly gift is only incurred if the amount of dough exceeds a certain measure [about 4 pounds].[52] Since this amount is fairly substantial, the Sages realized it would be tempting to divide the dough into portions weighing less than this minimum, and hence, exempt from the dough offering. This circumvention was regarded by the Sages with disapproval, and they prohibited it,[53] but did not rule that, if carried out, the evasive division would be ineffective.

Why did the Sages prohibit this evasion but not the others? There is no simple answer. One explanation is that the Sages wanted to prevent the law of the dough offering from falling into oblivion.[54] But it is not explained why this particular evasion would cause a law to be forgotten while the evasion of tithes, priestly gifts and second-tithes does not arouse such apprehension. It is conceivable that while the dough offering is easily evaded by an action not readily perceived as evasion, this is not the case with regard to the other evasions, and hence, they are permitted, while evading the dough offering is not. But in my opinion, the ease of evading tithes and priestly gifts by abandoning produce and then reclaiming it is comparable to that of evading the dough offering. In the case of abandonment of priestly gifts, the Sages issued an

50 Maimonides, *loc. cit.*

51 Rashi, *loc. cit.*, according to one interpretation.

52 Sifre Zuta, Shelah; Mishnah Eduyot 1:2; bEruvin 83a–b; bPesahim 48a; Mishnah Hala 4:3.

53 bMenahot 67a–b.

54 See Meiri, Hala, ch. 2, interpreting the second Mishnah.

enactment to change the law, but for some reason they did not similarly amend the laws of the dough offering. It is likely that the Sages were willing to make changes in the laws of abandonment on the basis of the principle "any abandonment (i.e., forfeiture) declared by the court is valid,"[55] granting courts the authority to confiscate property and declare it ownerless. This would surely cover enactments qualifying the law of abandonment itself. With regard to the dough offering, however, the law says that no dough offering should be taken from dough made from less than the specified minimum of flour. For the sake of preventing a few cases of evasion, the Sages did not want to introduce a radical change in this ritual law to the effect that, contrary to biblical law,[56] the duty of the dough offering would be incurred on any quantity of dough. The rule that small amounts of dough are exempt is directly derived from the Bible. One who takes the dough offering from dough of less than the minimal amount performs an action which is null and void, and the dough does not acquire holiness.[57] It is thus understandable that this type of evasion, although forbidden, is nonetheless valid.

This constitutes further evidence for our argument that as long as the Sages left a law unchanged, then even if they disapproved of an action, if it was carried out in a technically unassailable manner, that is, according to the letter of the law, they refrained from invalidating it.

vi. Selling Dough to a Non-Jew

As we saw, to prevent alleged – or actual – evasion by *baalei kisin*,[58] an amendment was enacted to the effect that the harvest of a non-Jew is also subject to tithes and priestly gifts. Does this amendment apply to the dough offering as well, since its laws are identical to those of the heave offering (*teruma*), another priestly gift?[59] The Sages originally argued that the two kinds of gift should not be differentiated, and thus, that the enactment should apply to the dough offering, but for various reasons, which we shall examine below, they ultimately distinguished between the two.

---

55 Tosefta Shekalim 1:1, bJebamot 89b, bGitin 36b.

56 Sifre Zuta, *loc. cit.*; bEruvin 83b.

57 Maimonides, *Code*, Laws Concerning First Fruits and Other Gifts to the Priesthood 6:16; *Shulhan Arukh*, Yore Deia 326:5.

58 See text at note 48.

59 See note 14.

In Tractate Menahot we find an attempt to apply the enactment to the dough offering.[60] This suggestion is countered by the argument that since the purpose of the enactment is to discourage evasion, there is no point in applying it to the dough offering, which can be evaded by simply baking less than the five measures of meal which constitute the minimal quantity subject to the dough offering. One who intends to bake a large amount of pastry can always evade the dough offering by dividing the dough into a number of batches each containing less than the minimum. On this view, it makes no sense to close one loophole by legislative enactment if others are still open. Indeed, the Gemara argues, there is no point in applying the enactment to tithes and to the other priestly gifts, which can also be evaded by other means."[61] As we saw, R. Oshea clearly states that they can be evaded by mixing the grain with chaff; we also saw that exemption can be effected by bringing the harvest home by way of rooftops and backyards.[62] Yet if it was possible to evade tithes and heave offerings by various stratagems, why was it necessary to rule that the crops of non-Jews were subject to these duties in order to prevent evasion by *baalei kisin*?

The Gemara concludes that there is indeed a distinction between the dough offering and the other gifts: "There, it is in public, and it would make him look stingy, here it is in private and it does not make him look stingy."[63] That is, one interested in doing so can evade all tithes and priestly gifts, and the enactment making non-Jews subject to these duties is ineffective. But any evasion of tithes and heave offerings has to be carried out in public. Even one who wants to evade will most likely be deterred from doing so. But since we are dealing with someone who wants to evade in order to save money, this person will not be deterred from secretive, non-public evasions, such as claiming a given crop is owned by non-Jews, or fictitiously transferring crops from Jew to non-Jew. Thus with regard to tithes and heave offerings the enactment makes good sense, since it may effectively eliminate the loophole of transactions with non-Jews, while the public and observable nature of other contemplated evasions may deter potential evaders.

With regard to the dough offering the situation is different. Eliminating one loophole by stipulating that the dough of non-Jews is liable would not be

60 bMenahot 67a–b.
61 See text at notes 15–20.
62 See text at note 25ff.
63 bMenahot 67b.

effective, since if someone wants to evade, this can be accomplished easily and discreetly by dividing the dough into smaller units. Hence there is no point in amending the law of the dough offering as was done in the case of tithes and heave offerings, and the law remains in its original biblical form – the dough of a non-Jew is exempt from the dough offering.

This discussion yields at least two important conclusions, the significance of which transcends the issues of priestly gifts and evasion: (1) The Sages were very cautious about making new laws or enactments and did so only if there was good reason to believe that the desired outcome would indeed result; and (2) In spite of the apparent indifference on the part of the Sages to evasion of tithes and priestly gifts – the evasion suggested by R. Oshea, for example, is not presented as morally problematic – both the Sages and ordinary individuals regarded evasion with disdain, and even those willing to evade in order to avoid tithes and gifts were ashamed to do so. The pericope indicates that public evasion was uncommon, and thus, it made good sense to eliminate the evasion arising from this loophole by making the crops of non-Jews subject to priestly gifts. It is evident that the Sages were uneasy about evasion of tithes and priestly gifts and tried to prevent it; since this was not possible within the existing law, they amended it. It is thus quite surprising to find that some of the Sages, such as Rabbi (Judah the Prince) and R. Jose b. R. Judah, were prepared to evade priestly gifts "through rooftops."[64]

---

64 See notes 27 and 30. The evaders are R. Judah bar Ilai's own son, R. Jose, and his student, Rabbi (Judah the Prince). This affinity may explain why he felt he could criticize and lecture them, contrasting their behavior with that of his own teacher, R. Akiba. The rest of the anecdote emphasizes the strangeness of their conduct: There was an old man with Rabbi and he asked, would you let me have it? Yes, they answered. He told them, You would not give it to your Heavenly Father, yet you give it to me?

R. Jose bar Judah is apparently consistent in his attitude to evasion, which he thinks is legally permitted. On two points related to the laws of the Sabbath he disagrees with the Sages and permits evasion (Tosefta Shabbat 13 (14):7; bShabbat 117b). Note that the name R. Jose b. R. Bun in the printed edition of jShabbat 16:3 should be changed to R. Jose bar Judah; see "the barrel that broke on the Sabbath," below. True to his views, R. Jose bar Judah also permits evasion of laws governing the festivals, see bMoed Katan 12b, and below, "Preparing food on a festival." However, it may also be possible to explain the anecdote in the Jerusalem Talmud (note 27 above) differently. Perhaps it concerned only a small amount of fruit in a basket, which Rabbi and R. Jose bar Judah considered a snack, to which different rules apply, and thus, they did not

A possible explanation is that the different attitudes of the Sages reflect different historical periods and prevailing customs.[65]

B. First-born Animals and Tithes on the Herd

We can also learn much about evasion from the laws concerning first-born animals and tithes on the herd. Here, unlike the laws of tithes and priestly gifts of produce, it is easier to understand why certain kinds of evasion were permitted, and how halakhically-condoned evasion evolved. There is also more direct evidence here as to the reasons for evasion and the Sages' tolerance of it.

i. First-born Animals

Firstlings constitute one of the twenty-four categories of gifts given to priests.[66] They belong to the category of *kodashim kalim* (gifts of lesser sanctity): after they are slaughtered and sacrificial portions are burned on the Temple altar, the rest may be eaten by the priests.[67] As is the case for all other

---

perceive it as intentional evasion. Support for this interpretation is found in the word *kalkala*, which denotes, according to definitions in both Jastrow and Ben Yehuda, a fruit basket or container small enough to be handled by one person; see Mishnah Peia 7:3; Tosefta Terumot 2:8. However, the *Shulhan Arukh* defines *kalkala* as a large container. In any event, the anecdote does not necessarily mean that Rabbi and R. Jose b. Judah intentionally evaded tithes on their produce. They may have used the basket since the quantity of fruit was not large.

65 See E.E. Urbach, "Megamot Datiot Vehevratiot Betorat Hatzedaka Shel Hazal," *Zion* 16 (1951), pp. 1, 11. In his "Hafrashat Maaser Rishon Halakha Lemaase Betekufat Bayit Sheini," A. Oppenheimer states that:

> The Sages themselves were aware of the difficulties involved in the scrupulous observance of the rules regarding tithes, and permitted, within the framework of the halakha, various relaxations and evasions of the laws, which in certain circumstances made it possible to avoid giving tithes. Most of these rules date from the post-Temple period, but some do date from the Temple period [*Sefer Zikaron Lebinyamin DeVries*, ed. E.Z. Melamed (Jerusalem: 1969), p. 82].

66 Tosefta Hala 2; bBaba Kama 110b; Maimonides, *Code*, Laws Concerning First Fruits and Other Gifts to the Priesthood 1:14.

67 Mishnah Zevahim 5:5; Maimonides, Laws Concerning Firstlings 1:2.

sacrifices, a firstling that is to be sacrificed must have no blemish. A blemished firstling loses its sanctity and may be used as an ordinary animal.[68] Unlike other *kodashim kalim*, the firstling does not have to be redeemed.[69] Since there is no longer a Temple where sacrifices can be offered, unblemished firstlings are sent to pasture until they become unfit for sacrificial purposes by being blemished through injury.

Hence, since the destruction of the Temple, a firstling is a useless creature. Since it must be fed but cannot be used or eaten, its owner incurs a financial loss. Like other animals dedicated for sacrificial use, it must not be shorn or used as a draft animal; it is thus a nuisance and a temptation.[70] Hence, after the destruction of the Temple, the Sages contemplated the possibility of evading the laws of firstlings altogether.[71]

---

68 Deut. 15:21–22; Sifre *ad loc.*

69 Mishnah Bekhorot 6:11.

70 Deut.15:19; Sifre *ad loc.*; Maimonides, *Code*, Laws Concerning Trespass 1:7. The verse from which the prohibition against shearing and otherwise using sanctified animals is derived reads, "Thou shalt do no work with the firstling of thine ox, nor shear the firstling of thy flock."

71 Intentional infliction of a blemish on a firstling was prohibited.

> If one wounds the ear of the firstling, it may never be slaughtered, according to R. Eliezer. The Sages say: When it gets another blemish it may be slaughtered because of that. And it happened there was an old ram which was a firstling, and its fleece was dangling; a Roman quaestor saw it and asked; What is wrong with this one then? The owner answered: It is a firstling and may not be slaughtered until it gets a blemish. The quaestor took a dagger and wounded the ram on the ear. The Sages were consulted and they permitted it. He saw that the Sages permitted it, he went and wounded other firstlings as well, but the Sages prohibited them. Once youths were playing in the fields and they bound together the tails of some firstling kids. The tail of one kid was torn. The Sages were consulted and they permitted it. They saw that it was permitted, and they tied together the tails of other firstlings, but the Sages prohibited them. This is the rule: Anything intentional is forbidden, if it is not intentional, it is permitted (Mishnah Bekhorot 5:3).

A similar form of evasion is mentioned in the Talmud concerning the castration of animals. On the basis of the anecdote related in bBaba Metzia 90a–b, Maimonides rules: "It is forbidden to tell a non-Jew to castrate our animals, but if he took it by himself and castrated it, it is permitted. If a Jew evades this law, he is to be fined by compelling him to sell this animal" (*Code*, Laws Concerning Forbidden Intercourse 16:13). Cf. *Zikhron Yehuda* #91 (45, 1).

(a) Dedication of the Unborn Animal as a Burnt Offering As early as the Second Temple period, when sacrifices were still being carried out, certain restricted kinds of evasion regarding firstlings were practiced. The Tannaim themselves suggested the method. Their motive is not clear. It is possible that they wanted to ease the financial burden on ordinary people who were obliged to offer sacrificial animals. Indeed, the Mishnah opens quite frankly: "How can one evade the law of the firstling?"[72] The proposed means of reducing expenses entails taking advantage of existing laws to reach the objective in question. The Sages suggest that the unborn firstling be dedicated as a burnt offering,[73] in which case the owner is not obliged to give it to the priests at birth, since it has already been promised elsewhere. Although he loses the animal, the owner does gain something: if he ever has to offer a burnt offering, the dedication of the firstling satisfies this obligation and he is not obliged to dedicate an additional animal. This evasion is built around the following two laws: (a) the dedication of an unborn animal is valid,[74] and (b) the sanctity of a firstling begins with its birth.[75] The dedication of the fetal firstling is an evasion, since ordinarily if one is obliged or intends to bring a burnt offering, he takes an animal from his herd, dedicates it, and brings it to the Temple. In the case before us, the owner dedicates an unborn animal, not because he is obliged to bring a burnt offering at the time, but to avoid the future expense of having to sacrifice two different animals should such an obligation arise. The Sages safeguarded the interests of the owner, rather than those of the priest, who, if the suggestion is heeded, forfeits the firstling due him.

On the basis of this evasion the Gemara suggests another possible evasion: dedication of the fetus of an animal previously dedicated as a sin offering as a sacrificial offering – say, a burnt offering – in its own right. This would allow use of the animal as a burnt offering after its birth and there would be no need to destroy it, in spite of the rule that the offspring of a sin offering must be destroyed.[76] However, this suggestion was rejected, as it was decided that the fetus, upon conception,[77] would bear the sacrificial status of the dam, with no

72 Mishnah Temura 5:1.

73 The Gemara concludes that a burnt offering is most appropriate. See below, "Inflicting a Blemish before Birth."

74 Mishnah, *loc. cit.*

75 bHulin 69b, Bekhorot 35a.

76 bTemura 24b; Rashi *ad loc.* s.v. *labeheima.*

77 bTemura 24b–25a.

possibility of changing this status in utero by rededicating the unborn animal.[78]

This Mishnah is important for two reasons. First, it undoubtedly reflects the situation during the Second Temple period, indicating a lenient attitude to, or even the tacit approval of, the practice of withholding firstlings from the priests. It may also reflect an attempt to encourage sacrificial offerings at the expense of the priestly gift of the firstling. Second, the Mishnah establishes a ruling of great significance for generations living after the destruction of the Temple: the sanctity of a firstling begins with its birth (i.e., with the emergence of its head or the greater part of its body); it has no sanctity whatsoever in the fetal stage. It is thus permissible, when the animal is in utero, to take steps to avoid the obligation to give it to the priest after birth.

(b) Inflicting a Blemish before Birth On the basis of this stratagem, the Sages developed two additional methods of evading the gift of the firstling: inflicting a blemish on it before the extrusion of its head or the greater part of its body; and its complete or partial sale to a non-Jew before its birth.[79] On the basis of the principle, derived from the Bible,[80] that a firstling may be dedicated while still a fetus, a solution was found for the problem of firstlings in the post-Temple period.

After the destruction of the Temple, it became impossible to evade the duty of firstlings by dedicating them as burnt offerings, and another solution had to be found, since the need for evasion was now far greater. Because sacrifices were defunct, priests could no longer make use of the firstling, yet were forbidden to intentionally inflict any blemish on it. Thus, not only does the firstling constitute a significant loss for its owner and subsequently for the priest who receives it and has to care for it though deriving no benefit from it, it also gives rise to another serious problem. The firstling must not be put to work or shorn; one may not benefit from it in any way. Violation of this

78 The Gemara's suggestion is based on a strict reading of the wording of the Mishnah, but it is doubtful whether an ordinary reading would support it. See Albeck's commentary on Mishnah Temura 5:1.

79 bBekhorot 3a–b.

80 bTemura 25a; Sifra, Behukotai 8 (p. 114, 3 in Weiss edition). The "not" (*ein*) should be deleted from Sifra's "the duty of the firstling may not be evaded", in light of the plain meaning of the biblical passage and the Gemara on bTemura. This emendation has been accepted by all the commentators, e.g., R. Elijah of Vilna; R.I. Di Trani on Sifra; Malbim in *Hatora Vehamitzva*.

prohibition by a firstling owner, whether unwitting or intentional, is punishable by divine extirpation (*karet*), yet very likely to occur, given the constant temptation having such an animal in his possession poses for the owner.

The Sages therefore sought a way for the law to be circumvented with relative ease. The solution is attributed to a Babylonian Amora of the second generation, R. Judah b. Ezekiel, the disciple of Rab and Samuel. His approach is simple: since the sanctity of the firstling commences upon the emergence of its head or the greater part of its body, it is permissible to disfigure it, or inflict a blemish on it, at the onset of parturition, before it emerges.[81] R. Judah's contemporaries were apparently reluctant to accept this novel opinion, for obvious reasons. Against him, the Gemara quotes the Mishnah discussed above[82] in an attempt to prove that the Tannaim were opposed, in principle, to any stratagem aimed at violating the firstling's sanctity. The Amoraim who oppose R. Judah contend that according to this Mishnah, it is forbidden to dedicate a firstling as a peace offering before its birth. Evasion is only allowed if the sanctity of the unborn firstling is increased by its dedication as a burnt offering. If a fetus cannot lose the firstling status it gains at birth in order to be dedicated as a peace offering, then certainly it must not be disfigured in order to negate its firstling status.

R. Judah himself agrees, arguing that the law may indeed have been more stringent in the Temple period, with dedication as a burnt offering the only means of evasion, as is recorded in the Mishnah. But now, in the post-Temple period, other avenues ought to be considered, since if firstlings were born, there would be a real likelihood they would be shorn or put to work.[83] R. Judah puts forward another novel argument. He says that disfigurement of an unborn firstling for the purpose of evasion was prohibited during the Temple period. This rule was not formulated previously and we have no direct evidence from the Tannaitic period to confirm it. But it seems R. Judah's account is correct, since during the Temple period there was another option for evasion – the animal's dedication as a burnt offering – though this option was less efficient from the owner's point of view, given that the firstling retained its sanctity, unlike a disfigured animal which had no special status. However, given that R. Judah was the one who suggested disfigurement

---

81 bBekhorot 3b, 35a, 53b; bTemura 24b.

82 bTemura, *ibid.*; the Mishnah referred to, Temura 5:1, is discussed above near note 72ff.

83 *Ibid.*

before birth, we have no way of knowing what the Tannaim would have thought of this solution when the Temple was still standing.

After R. Judah proposed his novel solution, it seems to have been put into practice, and in the fourth generation of Amoraim we find Rava explaining in detail where and how the disfigurement should be inflicted.[84]

(c) Sale to a Non-Jew During the same period, that of the fourth generation Amoraim, R. Mari bar Rachel suggested another novel way of evading the law of the firstling. In spite of initial resistance to his proposal, his view was ultimately accepted. After the talmudic period it greatly influenced the treatment of firstlings and the strategy adopted for evading the law of the firstling. Yet the commentators disagree over the exact nature of R. Mari b. Rachel's proposal.

They do agree that R. Mari b. Rachel indeed suggested that the animal, or part of it, be sold to a non-Jew before it assumed the sanctity of the firstling.[85] Whether this novel form of evasion is connected in any way to the origins of R. Mari b. Rachel is an interesting question. His father was a convert to Judaism and he himself had been conceived before his father's conversion, which explains why he is named after his mother, Rachel,[86] rather than his father. But this point has no direct bearing on our subject.

From the legal point of view, there is a similarity between the response to the innovation of R. Judah and the response to that of R. Mari b. Rachel. As we noted, R. Judah b. Ezekiel's view was initially attacked, but subsequently gained acceptance.

When R. Mari b. Rachel advanced his solution, he too was criticized. His opponents based their attack on the arguments of R. Judah, whose view had become the consensus. However, R. Mari b. Rachel's proposal, like that of R. Judah, was ultimately accepted. The passage in the Gemara is as follows:

> R. Mari b. Rachel used to have animals. He would sell their ears to a non-Jew and forbid shearing them or using them for any work and handed them over to

84 bBekhorot 35a.

85 bBekhorot 3b. All the passages in question, but this one in particular, clearly suggest that the Sages did not consider relevant the distinction between evasion by a physical act such as disfigurement of a firstling, and evasion by a legal act such as selling a firstling to a non-Jew.

86 For an extensive discussion of whether there were one or two Amoraim named R. Mari, see A. Hyman, "Rav Mari Bar Isur Giora Veimo Rahel Bat Shmuel," in *Toldot Tanaim Veamoraim* (Jerusalem: 1964), part 3.

> the priests. The animals of R. Mari b. Rachel perished. If he forbade their shearing and did not use them for any work and gave them to the priest why did he also sell their ear to a non-Jew? He wanted to ensure himself against mishap. If so, why did his flock perish? Because he annulled their sanctity. But R. Judah said: Is one permitted to disfigure a firstling before its birth? There, he annulled the sanctity of fitness for the altar but the sanctity to be given to a priest was not annulled. In this case even the sanctity of fitness for a priest has been annulled. And it might also be that R. Mari b. Rachel knew how to perform a proper legal transaction, but others might imitate him, thinking that R. Mari only talked to the non-Jew, and a mishap might result.[87]

Upon examination, the passage indicates that the Talmud prefers disfigurement to sale, since the blemished animal retains its firstling sanctity and remains the property of the priest, losing only its fitness as a sacrificial animal, while sale to a non-Jew annuls its sanctity altogether. This distinction was not fully accepted by the editor of the pericope, who added the second rationalization for the punishment of R. Mari b. Rachel.

Why was this second argument, implying that the first solution was unsatisfactory, introduced? The answer appears to be in the Jerusalem Talmud,[88] where we find a strong attack on the suggestion that sales to non-Jews, which are perfectly legitimate, are unacceptable, because transferring the merchandise from Jew to non-Jew has repercussions as to the religious obligations connected with it which would have obtained had it stayed in the possession of a Jew. The Jerusalem Talmud presents this as a reductio ad absurdum – is it forbidden to sell wheat to a non-Jew since in the future no dough offering will be set aside from it? Should we not sell wine and oil to non-Jews since if we do blessings will not be made before their consumption?

The editor of the pericope, aware of this argument, apparently felt obliged to bring a second answer, beginning "It may also be said." Its interpretation is debated in the commentaries.[89] According to this explanation, R. Mari b. Rachel performed an action that was permissible in principle, but problematic in terms of the way in which it was carried out. The transaction might not be understood by laymen, who might fail (i.e., act impermissibly) if they try to follow his example. As a result, though the action itself is acceptable, and

87 bBekhorot 3b.

88 jAvoda Zara 1:6 (39d).

89 See, e.g., *Tosafot*, Bekhorot 3b s.v. *deka mifka lahu*; *Sefer Hayashar Lerabeinu Tam*, Helek Hahidushim, 521.

there is no reason to object to the sale per se, his method of evasion is forbidden. In view of its public nature, problems might arise if it were copied.

After the method of R. Mari b. Rachel was introduced, however, the halakhic authorities decided that sale of the animal, or parts of it, to a non-Jew was the preferred means of evading the duty of the firstling. The practice of disfigurement before birth was discontinued altogether after the talmudic period, and selling the dam before she gave birth became the accepted method of evasion.[90] This method resembles that of R. Mari b. Rachel, but differs on one important point: R. Mari b. Rachel apparently sold the firstling itself, not its mother. However, post-talmudic developments are beyond the scope of this discussion.

ii. Tithes on the Herd

The law states that every tenth animal from herds of cattle and flocks of sheep must be set aside.[91] This animal, the tithe on the herd, gains a sacrificial status similar to that of the "firstling." That is, it falls under the category of *kodashim kalim* and is not given to the priests as a gift. Rather, its blood is sprinkled on the Temple altar and its fat burned; the rest is consumed by its owners in Jerusalem.[92]

As we noted, sacrificial rites ceased with the destruction of the Temple. However, the legal obligations concerning tithes on the herd still obtain. The Mishnah states: "Tithes on the herd obtain in and outside the land of Israel, whether or not the Temple exists."[93] Thus, though the obligation is still binding,

90 See Tosefta Bekhorot, *loc. cit.*; Asheri, *Responsa*, #49, 2; Tur, Yore Deia 320:9; Shulhan Arukh, Yore Deia 320:6. In another pericope, the status of one who is half-slave and half-free is considered. According to the law, he must be freed. The half-slave's owner transferred ownership to his minor son. The Sages could not annul the transfer since all the legalities had been observed, and the minor could not be compelled to set his half-slave free. This case is different from that of the transfer of ownership of the firstling to a non-Jew, since the duty to set the slave free becomes binding on the son when he reaches adulthood, but the duty of the firstling is avoided altogether in the transaction with the non-Jew. However, the Sages found a solution to the problem; see bGitin 40a–b.

91 See the ninth chapter of Mishnah Bekhorot.

92 For a tradition regarding the tithe on the herd which was not part of the original law, namely, the tradition that it was given directly to the priest rather than sacrificed, see "Terumot and Ma'aserot," *Encyclopedia Judaica*, vol. 15, pp. 1025–1026.

93 Mishnah Bekhorot 9:1.

there is no way to discharge it. The problem is similar to that posed by the firstling. There is an illuminating discussion in tractate Bekhorot on some of the issues this raises, including that of evasion.

It begins with a debate on the Mishnah's statement that the tithe on the herd is due "whether or not the Temple exists."[94] An anonymous Amora asks, "Do we, according to this Mishnah, have to tithe our herds?" Implicit here is the argument that it is well-established that herds are not, in fact, tithed. How can the Mishnah be reconciled with custom? The editor offers a solution in the name of R. Huna, but it is rejected. Then Rava (not Rabba as in some manuscripts) says that the Sages issued a decree (*gezeira*) to prevent mishap. Their apprehension here is the same as in the case of the firstling. Because the animal cannot be shorn or used for work, it would be a source of constant temptation. In support, a *baraita* is quoted: "In our times animals are not dedicated as holy, neither they nor their worth." Apparently, tithing falls into the category of "dedicated as holy."

The questioning continues: If so, why does the obligation of the firstborn still apply? The answer is that the birth of an animal is not under our control. As soon as a firstborn emerges, it is holy, and an enactment against nature is of no avail. But tithes are different. As long as the tenth animal has not been set aside, it is not holy. Given the enactment against tithing, there is no longer a tithe on the flock. While, in view of the fact that it overrides a biblical precept, the enactment is highly significant, it should be noted that compliance with it is secured by refraining from action (*shev veal taase*).

The discussion then turns to the question of evasion. The Gemara explains an earlier statement: there is no need for an enactment regarding firstlings because there is a simple way to prevent what would be the birth of a sanctified firstling: selling part of the fetus or its mother to a non-Jew.[95] The Gemara then suggests another solution to the problem, one which is already familiar to us. It implies that this solution is preferable to that of selling to a non-Jew. This second solution is the suggestion of R. Judah – infliction of a blemish to render the fetus unfit for holy service. In my opinion, the pericope was compiled after the suggestion of R. Mari b. Rachel had been proffered, and that suggestion was, indeed, taken into account, and even vetted as a potential solution. It is clear that the pericope's editor preferred the earlier

94 bBekhorot 53a–b.

95 On whether the animal itself, or its mother, is sold, see the various commentaries on the pericope, and the sources cited in note 89 above.

solution of R. Judah (disfigurement) to the later solution of R. Mari b. Rachel (sale to a non-Jew). This accords with what we concluded above.[96]

Next, the Talmud asks why tithes on the herd were abrogated. Would it not have been possible to maintain the law but circumvent its consequences by inflicting a blemish, as suggested in Bekhorot? After considerable discussion, the conclusion is that the only way to implement the suggestion would be to disfigure the whole herd, since it could not be predicted in advance which animal would be tenth when the animals were actually counted. However, since the rebuilding of the Temple is always a possibility, in which case, if all animals were blemished, there would be a shortage of fit sacrificial animals, the Gemara decides that this method of evasion must be rejected.[97]

iii. Summary

It may be concluded that: (a) It is permissible to inflict a blemish on a firstling but not on the whole herd or flock from which the tithe is to be taken; (b) It follows that disfigurement is preferable to sale to a non-Jew; (c) From the conclusion of the discussion, we see that the tithe on the herd is no longer extant due to fear that the holy animal would be put to use. A suggested remedy – disfigurement of entire herds – is deemed unsuitable, because it would lead to a dearth of fit sacrificial animals; (d) The law of firstlings is still applicable in spite of the destruction of the Temple, however, because even if a blemish were to be inflicted on firstlings, there would be other fit animals available for sacrifices. In other words, since there is a way to avoid the law of the firstling without threatening the availability of sacrificial animals, the laws of the firstling still apply. Nevertheless, it is desirable to circumvent the obligation by some means. But the case of tithes is different, since if all the animals were to be blemished, should the Temple be rebuilt, there might be a shortage of fit sacrificial animals for the Temple service. So the Sages issued an enactment to the effect that the tithe on the herd no longer applies. That is, they rescinded the biblical command of tithing the herd to guarantee immediate availability of sacrificial animals when the Temple is

---

96 See "Sale to a Non-Jew," above.

97 Literally, "Soon the Temple will be rebuilt and we will need sacrificial animals and there will be none" (bBekhorot 53b). The notion appears elsewhere as well. Identical in phrasing, it is applied to the question of priests drinking wine after the destruction of the Temple: "Soon the Temple will be rebuilt and we will need priests fit to perform the services and there will be none" (bTaanit 17b).

rebuilt. Otherwise, people would have blemished all their animals to avoid the temptations posed by such animals.

C. The Leper's Guilt Offering

As part of their purification rites, lepers are obliged to bring a sin offering.[98] By law, if an animal is dedicated as a sin offering, it must be put to death should its owners die.[99] The people of Alexandria queried R. Joshua about a case to which these laws pertain: there are two lepers, each of whom has dedicated an animal as a sin offering. The two animals were mixed up, so that it is not known which animal belongs to which owner. One of the animals is sacrificed and one of the lepers dies. A serious legal problem exists, in that a sin offering must not be sacrificed, that is, an animal must not be slaughtered, if there is any doubt as to whether it is due.[100] It would not resolve the problem to advise the leper to bring a fowl as a sin offering, since though fowls may be sacrificed even if an element of uncertainty exists, only a pauper may use a fowl as a sin offering, and a rich man does not discharge his obligation with a fowl.[101] The surviving leper is not a pauper. What is he to do?

R. Joshua[102] replied to the people of Alexandria that the leper should transfer his property to someone else – someone he trusts to return it later – and become a pauper; he will thus be entitled to bring a fowl as a sin offering.[103]

As this example shows, when the Sages sought a satisfactory solution to a legal problem, they were willing to recommend actions that clearly involved evasion. Here, as we saw with regard to tithes,[104] and firstlings,[105] and as will be the case when we examine the sale of leavened bread before Passover,[106]

---

98 See Lev. 14:10–19.

99 See Mishnah Temura 4:1, Mishnah Meila 3:1, bMeila 10b.

100 Tosefta Temura 4:20, Tosefta Nezirut 6:1.

101 Mishnah Negaim 14:12, jNazir 1:2 (57b).

102 Apparently, R. Joshua had a tendency to be lenient on matters of evasion. See below, "Extracting an animal from a pit… ."

103 Mishnah Negaim 14:13; Tosefta Negaim 9:9; cf. Tosefta Nezirut 6:1; bNida 70a; jNazir 1:2 (57b). In Tosefta Negaim, this comment is appended to R. Joshua's reply to the people of Alexandria: "Now you have asked a thing of wisdom." See S. Lieberman, *Tosefta Kifshuta* (New York: 1955), Zeraim, vol. 1, p. 47.

104 See above, "Transferring Ownership."

105 See above, "Sale to a Non-Jew."

106 See below, "Leaven on Passover."

the condoned evasion involves the fictitious transfer of property to avoid legal obligations. Though R. Joshua does not explicitly use the term "evasion," there is no doubt that this constitutes a clear case of recommending an evasive course of action.[107]

D. Vows

An area in which questions related to evasion arise relatively frequently is that of the laws pertaining to vows. Overall, the halakha takes a dim view of vows and oaths, for they impose unnecessary restrictions. The general attitude is that, where possible, release from vows is desirable.[108] This is not to say that vows can always be annulled without reservation, but rather, that dissolution of vows is sanctioned if carried out in compliance with accepted halakhic criteria. It is thus not surprising that if there is a way to circumvent a vow by evasion of the law, it is condoned. Nevertheless, even with regard to vows, certain constraints remain in effect and not every form of evasion is acceptable.

There is an incident described in the Bible which closely parallels the Sages' attitudes to evasion in general, and evasion of vows in particular. It is astonishing that this was not pointed out in the talmudic literature in support of the Sages' rulings. A satisfactory explanation has yet to be found. I refer to the last chapter of Judges, after the incident of the concubine at Gibeah and the oath that "There shall not any of us give his daughter unto Benjamin to wife." The passage is as follows:

> And the people repented them for Benjamin, because that the Lord had made a breach in the tribes of Israel. Then the elders of the congregation said: How shall we do for wives for them that remain, seeing the women are destroyed out of

107 I thank Berachyahu Lifshitz for drawing the case of the leper's sacrifice to my attention.

108 There are many examples of this, see e.g., bNedarim 21b–22a, 23a, 59a, and particularly, ch. 9 of Mishnah Nedarim. Cf. Z. Falk, "Binding and Loosening," *Journal of Jewish Studies* 25 (1974), pp. 92–100. Biases on the part of commentators are often reflected in their views on the question of evasion; matters related to vows constitute fertile ground for discussion of this point since many of the suggested remedies are based on alternative ways of interpreting the vow-maker's words; often, such interpretations are quite far-fetched. The greater part of Nedarim deals with these matters.

> Benjamin? And they said: They that are escaped must be as an inheritance for Benjamin, that a tribe be not blotted out from Israel. Howbeit we may not give them wives of our daughters. For the children of Israel had sworn, saying: Cursed be he that giveth a wife to Benjamin. And they said: Behold, there is the feast of the Lord from year to year in Shiloh, which is on the north of Bethel, on the east side of the highway that goeth up from Bethel to Shechem, and on the south of Lebonah. And they commanded the children of Benjamin, saying: Go and lie in wait in the vineyards; and see, and behold if the daughters of Shiloh come out to dance in the dances, then come ye out of the vineyards, and catch you every man his wife of the daughters of Shiloh, and go to the land of Benjamin. And it shall be, when their fathers or their brethren come to strive with us, that we will say unto them: Grant them graciously unto us; because we took not for each man of them his wife in battle; neither did ye give them unto them, that ye should now be guilty. And the children of Benjamin did so, and took them wives, according to their number, of them that danced, whom they carried off; and they went and returned unto their inheritance, and built the cities and dwelt in them.[109]

The Israelites had sworn not to give any of their daughters in marriage to the Benjaminites, but regretted this decision and sought to rescind their oath. Their motive was laudable – they wanted to prevent the extinction of the tribe of Benjamin. So they circumvented their oath: they did not **give** any of their daughters to the Benjaminites, but caused the Benjaminites to take them themselves without previously obtaining the girls' fathers' consent. In this story, we find all the characteristic features of the rescinding of vows, as the concept was understood at the end of the talmudic period. As noted, however, the Talmud does not mention this ancient story as a historical precedent for evasion or for rescinding vows.

In the Mishnah and the Gemara, a number of similar strategies for evading vows are mentioned. Several examples in Nedarim involve cases in which someone vows not to benefit from someone else, while the latter wishes to bestow benefit on the vow-maker – that is, to do that which is forbidden by the vow. The problem is similar to that recounted in Judges. In the Talmud, the solution to the problem involves a third party, to whom the vow-maker hints that he would be interested in having him perform the forbidden [benefit-bestowing] act. An overt request does not solve the problem, because it turns the third party into an agent of the vower, and hence, forbidden as

---

109 Judges 21:15–23; see Malbim and *Meshekh Hokhma* commentaries *ad loc.*; cf. bBaba Batra 121a.

well, since one's agent is like oneself. We find the following passage in the Mishnah:

> One vowed not to benefit from his fellow but has nothing to eat. His fellow goes to the shopkeeper and says: So-and-so vowed not to benefit from me and I do not know what to do. He [the shopkeeper] gives him and takes from the other. The one who vowed had a house to build, or a fence to erect, or a field to harvest, his fellow goes to the laborers and says: So-and-so vowed not to benefit from me and I do not know what to do. They perform the task and take their wages from him.[110]

We will not go into the fascinating questions that arise from this Mishnah – such as whether "having nothing to eat"[111] is a precondition, or whether the third party can claim his expenses from the vow-maker – since they are not directly related to our concerns.[112] What is clear is that it is sometimes possible to evade a vow by hinting to a third party that he ought to perform an action forbidden to the hinter.

There is similar example in Mishnah Ketubot. "One who vows that his wife should have no benefit from him appoints an administrator."[113] The Gemara points out that if, as the Mishnah seems to say, an administrator is appointed, he is an agent of the husband and legally his actions are as those of the husband. Hence, they would constitute a breach of the vow. The Gemara suggests that the Mishnah should be understood in the spirit of Mishnah Nedarim: "R. Huna said, The husband says, Those who look after her will not lose."[114]

Evasion may be effected, not just by word or hint to another, but also by a combination of word and action. However, the vow-maker must be careful not to speak too directly. Through a third party, one may give a gift which is really intended for the "forbidden" person [the one upon whom benefit

110 Mishnah Nedarim 4:7.

111 Cf. *Tosafot* Nedarim 43b s.v. *noten leaher; Shita Mekubetzet* and *Peirush Harosh ad loc.*; *Tosafot* Nedarim 43a s.v. *hamudar hanaa mihaveiro; Peirush Harosh ad loc.; Piskei Harosh*, Nedarim, ch. 4, sec. 10; *Korban Netanel ad loc.*, sec. 8; Rashash, *ad loc.* s.v. *haya beito; Shita Mekubetzet, ad loc.* s.v. *veein lo mah yokhal.*

112 See jNedarim 4:7 (38d); Ran, Nedarim 43a s.v. *uba venotel mize; Peirush Harosh ad loc.* s.v. *hu noten.*

113 Mishnah Ketubot 7:1.

114 bKetubot 70b–71a.

cannot be bestowed], but this must not be stated explicitly. Care must be taken to avoid the classic error illustrated by the case of Beth Horon:

> It happened with one at Beth Horon that his father was forbidden by vow to gain benefit from him. This person was about to celebrate his son's marriage, and said to one of his friends: The yard and the food are all yours but only if father comes and participates in the festivities. The friend answered: If it is mine let it be all dedicated to the Sanctuary. He said: I did not give you these things to be dedicated. He answered: You only gave me these things so that you and your father could eat and drink together and be reconciled, while the sin would be mine. When the matter was brought before the Sages, they said: If a gift cannot be dedicated to the Sanctuary, it is not a gift.[115]

Here, in contrast to similar actions which were condoned, the vow-maker's action is condemned, though all these actions have a common goal, namely, to enable the forbidden person to benefit from the property of the vow-maker. The action here is seen as defective, because, to quote Silberg,

> He restricted the legal effectiveness of the gift in advance, so that it did not obtain under all circumstances, but only for the benefit of his father. Now a gift that cannot be used without reservation is no gift at all... the gift at Beth Horon was declared void because its scope was restricted.[116]

The gift at Beth Horon became a halakhic prototype. It is cited a number of times in the Mishnah, sometimes as a brief reference, sometimes in an attempt to examine its significance and compare it with other cases.[117] From our point of view, what is important is that it shows that evasion is possible, if carried out discreetly, since the evader's flaunting his true intentions may well subvert the evasion.

Further evidence of evasion of such vows by fictitious transfer of ownership is found in another Mishnah, which relates that it was customary to transfer property to a third party so that a forbidden person might benefit.[118] In special cases, property was transferred to "the president";[119] while the

115 Mishnah Nedarim 5:6.

116 M. Silberg, *Talmudic Law and the Modern State*, trans. B.Z. Bokser (New York: 1973), p. 31.

117 bKetubot 71a; Nedarim 43b, 48a–b; cf. bBaba Batra 133a–134a.

118 Mishnah Nedarim 5:6.

119 Mishnah Nedarim 5:5.

interpretation of the phrase is controversial, this has no bearing on the point under discussion.[120]

Another means of evasion, though similar to the methods considered above, is somewhat more problematic. There is no consensus regarding this method, also cited in Nedarim: "They are on the way and he has nothing to eat. He gives food to another as a gift and he may enjoy it."[121]

The Mishnah then considers a case in which there is no third party:

> If there is no one else with them he puts it on a rock or on a wall and says: this is abandoned and open to anyone who wishes to take it; the other takes it and eats. R. Jose forbids this.

It is not hard to understand the rationales behind these positions. According to the Sages, the abandonment of the property removes it from the possession of the vow-maker, so the forbidden person can take advantage of the situation without benefitting from the vow-maker. R. Jose argues, however, that if the two are alone it clearly means that the vow-maker gives his property to the forbidden person. Abandonment of property in the presence of someone known to be interested in taking possession of it is tantamount to giving it to him directly. At first, both positions appear quite tenable and it is not difficult to understand why, in this particular case, and **only** in this case, R. Jose differs with the other Tannaim. But according to Rava, the reason R. Jose is less lenient is "to avoid the deed of Beth Horon."[122]

R. Johanan offers a different explanation for R. Jose's position. He argues that R. Jose equates abandonment with giving a gift: "a gift is the property of the donor as long as the beneficiary has not taken possession of it, so it is with abandoned property, which belongs to its previous owner until someone takes possession of it."[123] Therefore, when the forbidden person takes the abandoned property, it is as if he takes it directly from the vow-maker. The Gemara appears to reject this approach.[124]

---

120 See H. Albeck's additional notes to his Mishnah commentary, (Jerusalem: 1959), Nedarim 5:5. According to Urbach and others, the Mishnah refers to the Hasmonean rulers; see E.E. Urbach, "*Hefker* – the Institution and its Role in History" (Hebrew), *Shenaton Hamishpat Haivri* 6–7 (1979–80), p. 19.

121 Mishnah Nedarim 4:8.

122 bKetubot 71a, bNedarim 43b.

123 *Ibid.*, 43a.

124 For the entire discussion, see bNedarim 43a–45a.

The approach to evasion of vows found in Nedarim is echoed in one of the accepted explanations of a passage, found elsewhere in the Talmud, concerning evasion of interest by means of a double sale. As is the case regarding vows, it is possible to evade interest by hinting, but open evasion such as that encountered in the Beth Horon case is not acceptable. If in the first transaction it is not mentioned that the buyer will resell the property to the vendor, the transaction is valid, though the parties concerned know very well that their intention is to evade the prohibition against taking interest. But if it is explicitly stated that there will be a double sale, then the transaction is defective, on the grounds that it evades the laws pertaining to interest, and hence, invalid.[125]

E. Leaven on Passover

The sale of leaven before Passover to a non-Jew is familiar to all. It originates in the mishnaic period, although it began to be practiced only later.[126] The Tosefta points in the direction of this solution in two consecutive passages. While they illuminate the positions taken by some of the Tannaim, their exact meaning is not unequivocal. In spite of the various interpretations that are possible, these passages nonetheless constitute sources for the sale of leaven to non-Jews to evade the prohibitions, "it shall not be seen," and "it shall not be found."

The first passage reads:

> A Jew and a non-Jew are sailing in a ship and the Jew has some leaven, he sells it to the non-Jew or gives it to him as a gift and takes it back after Passover, but it has to be given away as a gift without reservation.[127]

The second, immediately following, is:

> A Jew is permitted to say to a non-Jew, Instead of buying for a *mane*, buy for two, because I might need it and buy it from you after Passover.[128]

---

125 See the novellae of Nachmanides, the Rashba, and *Nimukei Yosef* on the Gemara's discussion about evading interest at bBaba Metzia 62b.

126 See S. Zevin, *Hamoadim Bahalakha* (Jerusalem: 1980), pp. 245ff.

127 Tosefta Pesahim 2:12; for textual variants of the two rulings, see Lieberman, *Tosefta Kifshuta, ad loc.*

128 Tosefta Pesahim 2:13.

These two laws, worded slightly differently, also appear consecutively in the Jerusalem Talmud,[129] but are not found in the Babylonian Talmud.

For our purposes, we have to bear in mind that the transfer must be in the form of a "gift without reservation." As we saw, the Tannaim made a similar stipulation concerning vows, and the gift at Beth Horon was deemed defective because it was not given without reservation.[130] Here, the exact meaning of a "gift made without reservation" is not provided by the Jerusalem Talmud, so various definitions of what constitutes "reservation" are possible, for example, a gift which must be returned,[131] or, a conditional gift.[132] The phrasing of the Tosefta suggests that the resale of the leaven to the Jew after Passover must not be an explicit condition of the sale, though it is permissible to implicitly convey that this is indeed the intention of the donor or vendor. It is also apparent from the Tosefta that this procedure is restricted to special cases, such as sea-voyages, and must not be used on an ongoing basis.[133]

The second law in the Tosefta is even less clear as to the precise nature of the transaction. A number of interpretations have been offered and it is difficult to say which of them reflects the original meaning of the Tosefta. The interpretation of Rabbenu Manoah may be the most plausible. "When you buy your supplies for the voyage for a hundred dinar, come and buy for two hundred for the requirements of both of us."[134] If this is the correct interpretation, then there is no evasion via the sale of leaven to a non-Jew, but rather, a request that the non-Jew buy a sufficient supply of leaven to enable the Jew to buy from him after Passover, when they will be at sea. But according to most of the Early Authorities, including an alternative explanation proposed by Rabbenu Manoah, the transaction described is as follows: the Jew demands an exorbitant price, so the non-Jew will have cause to change his mind and sell the leaven back; at the time of the first sale, the Jew hints that

---

129 jPesahim 2:2 (28d).

130 See the discussion in the previous section.

131 E.g., *Rokeah*, sec. 268; *Maase Rokeah*, Laws Concerning Searching for Leaven, sec. 21, p. 8; *Hagahot Maimoniot*, Laws Concerning the Sabbath 6b; cf. Radbaz, *Leshonot Harambam*, sec. 43; Lieberman, *op. cit.*, p. 495.

132 Maimonides, *Code*, Laws Concerning Leavened and Unleavened Bread 4:6; Lieberman, *op. cit.*, pp. 495–6.

133 See Maharam Halawa, Pesahim 30b s.v. *akum*; attributed to Ritba, Pesahim, beginning of ch. 2 (p. 5c); Meiri, Pesahim 6:1 (p. 66b).

134 *Sefer Manoah on the Rambam*, Laws Concerning Leavened and Unleavened Bread 4:6 (11b–12a).

he will buy back the leaven at the same price after Passover.[135] This way the leaven belongs to the non-Jew during Passover, but there is an understanding, though not an explicit condition, that the Jew will buy the leaven back. According to a number of variant readings quoted by the Early Authorities,[136] the prices given in the Tosefta are reversed, i.e., the Jew sells for one *mane* but hints that he is willing to buy it back for two. Apparently, though, the first version is more widely accepted.

Yet another method of evasion from the talmudic period is abandoning ownership of leaven on Passover and reclaiming it afterwards. This method, too, is found only in the Jerusalem Talmud, and not the Babylonian. In the Jerusalem Talmud, there is a dispute between R. Johanan and Resh Lakish over this method: R. Johanan objects to it, and Resh Lakish condones it. Their positions were not clear to later Amoraim, who settled the dispute as follows: "R. Johanan suspects evasion and Resh Lakish does not suspect evasion."[137] It is not stated what kind of evasion is suspected. Various explanations have been suggested, such as that of the *Korban Haeida*: "Perhaps he did not abandon it but just said that he did."[138] If this is indeed the meaning, then there is no evasion here, but fraud. On the other hand, if, as the *Pnei Moshe* commentary says, the evasion is simply the abandonment of leaven before Passover with the intention of claiming it after the festival,[139] then the matter is indeed one of evasion proper.

To summarize, evasion by sale is permissible under special circumstances. Evasion by abandonment depends on how the argument between R. Johanan and Resh Lakish is settled. It is not clear whether their disagreement is over the question of evasion. Even if we suppose that it is, it is not clear whether they define evasion as we do, or whether the argument is really over what we would call fraud.

F. Marital Law

This section will examine instances of evasion in which, to circumvent a particular law, an action ordinarily undertaken to achieve a certain end is

---

135 See Maimonides, *loc. cit.*; Lieberman, *op. cit.*, pp. 496–7.

136 *Sekhel Tov Shemot* 21:12, p. 105; cf. Lieberman, *loc. cit.*

137 jPesahim 2:2 (28d).

138 *Korban Haeida ad loc.* s.v. *ela Rav Yohanan.*

139 *Pnei Moshe ad loc.* s.v. *eina ken.*

carried out in order to achieve another. As far as I can determine, in all these cases the evasion is legally valid. If the purpose of the evasion is commendable, the Sages' attitude to it is likewise positive, but if the evader's aim is undesirable, the Sages regard the evasion as a negative act, though legally valid, since it is performed in accordance with recognized law.

The first such evasion we will consider is an act of R. Tarfon's, related in the Tosefta as follows:

> R. Menahem b. Napah said in the name of R. Eliezer Hakapar: It happened that R. Tarfon betrothed three hundred women and fed them on the priestly heave offering because it was a year of drought.

The Jerusalem Talmud calls this evasion:

> What is evasion? Did not R. Tarfon the patriarch of all Israel carry out an act of evasion? He betrothed three hundred women in a year of famine in order to feed them on the priestly heave offering.[140]

Strictly speaking, R. Tarfon evaded the intent of the relevant laws. In accordance with biblical law, women betrothed to priests can partake of the heave offering (*teruma*). By an enactment of the Sages, this privilege is not enjoyed immediately upon betrothal, but only after twelve months, or when the wedding ceremony takes place.[141] It is likely that the betrothal of the three hundred women by R. Tarfon took place specifically so that they would be able to eat of the heave offering immediately, or that, because of the famine, R. Tarfon waived the enactment and reverted to biblical law.[142]

In the ordinary sense the act in question is an evasion of the intent of the legislator.[143] Betrothal is the first step toward complete conjugal life and the law regards the betrothed woman as married in almost every sense. Biblical law allows the betrothed woman to partake of the heave offering if her betrothed is a priest, since she is soon going to live with him. In the case of R. Tarfon, however, the situation is different. He had no intention of marrying these women, hence, no intention of bringing them to his home. The intent of

---

140 Tosefta Ketubot 5:1, jJebamot 4:12 (6b).

141 Mishnah Eduyot 7:9, bNida 44b, bGitin 58a, bKetubot 47b.

142 See H. Albeck, "Haeirusin Veshtarotehem," in *Kovetz Madai Lezekher Moshe Shor* (New York: 1945).

143 See Silberg, *op. cit.* (note 116), p. 168, note 7.

the legislator which was to enable a priest's wife to eat together with her husband, was deliberately thwarted here. On the face of it, the halakha should have objected to such an action. However, since, by law, the intentions of those involved are not scrutinized at the time of betrothal, then even had it not been a year of drought, though there was no intention to go through with the marriage, the betrothal is legally valid, and the women may eat from the heave offering. Furthermore, if a marriage is undertaken for a purpose which is alien to the usual sense of marriage, for example, if its purpose is solely to enable a woman to eat from the heave offering, the betrothal and marriage are still legally valid.[144]

It has been noted that the Tosefta and the Jerusalem Talmud not only uphold the validity of R. Tarfon's action, but do so in a manner indicating that his conduct was not merely legal, but even commendable. In this respect R. Tarfon's action is different from similar actions which lack a meritorious objective.

Conceptually, R. Tarfon's action can be compared to the case related in the Gemara where a man married a woman because he needed storage space for his wine vats. Previously, the woman had been unwilling to rent him the required space, so he betrothed her, stored his property, and promptly divorced her. In spite of their negative attitude to this betrothal, and even more so, to the divorce, the Sages did not even consider the possibility that the betrothal itself might not be valid due to the groom's ulterior motive.[145] Nor is a betrothal void because it is intended as a means of gaining access to a wealthy woman's property.

A comparable case of evasion, in my opinion, is that of R. Huna, the son of Moshe bar Otzri. R. Huna was a young scholar, married and impoverished. His wealthy father, Moshe bar Otzri, was the guarantor for his marriage contract (*ketuba*). R. Huna had tried to get financial support from his father, apparently without success. Abbaye advised R. Huna to divorce his wife and thus compel his father to pay the sum settled on her by the marriage contract he had guaranteed. The couple would then be able to remarry and live off Bar Otzri's money. In the end this scheme came to nothing, because R. Huna was of priestly lineage, and thus would have been unable to remarry his wife if he divorced her. But the Gemara asks how the great Abbaye could have advised R. Huna on such a dishonest and devious course of action for the sake of

144 Similar issues arise in the context of present-day fictitious marriages; see *ibid.*, pp. 26–27.

145 bBaba Metzia 101b.

gaining the guarantor's pledge. The Gemara contends that the case is exceptional, for two reasons: a father and son were involved, and R. Huna was a scholar.[146] Had R. Huna been a simple layman rather than a promising scholar, or had the guarantor not been his father, Abbaye would not have given the same advice. Nevertheless, the Gemara accepts the scheme as legally valid, though it calls one who gives such advice a "cunning wicked man."[147] In any case, we see that it is indeed legal to get a divorce so that laws pertaining to the divorcee become applicable, though both parties know the divorce is a fiction and plan to remarry as soon as possible. In principle, the divorce, with all its legal consequences, is valid. The rationale behind Abbaye's advice here, apparently, is that a rich person ought to support his son and his son's family if the son has no means of his own yet wants to study.[148]

In both the case of R. Tarfon and that of R. Huna, the aim of the evasion was positive. R. Tarfon intended to feed starving women, while in the case of R. Huna, Abbaye sought a way to compel a parent to support his son's studies. In each case the course of action recommended or taken was legally unassailable. R. Tarfon betrothed the three hundred women quite legally, and R. Huna would have had a perfectly legal divorce. The consequences of both acts would have been those that follow from the acts by law. In neither case are the Sages censorious, since the goals of the evasions are desirable. As is clear from the story of Bar Otzri, had the goals been undesirable – for example, if the guarantee had been that of an innocent third party, and the beneficiary not a scholar – the acts and their consequences would still have been legal, but the Sages would have been uncomfortable with them, deploring them as acts of "cunning, wicked men."

3. The Second Type of Evasion: Undeclared Intentions and False Appearances

A. The Sabbath and Festivals

In this section we will examine cases in which the action per se is not decisive, but rather, it is the evader's subjective attitude that requires scrutiny.

146 bBaba Batra 174b; bArakhin 23a–b; see the commentaries on whether the two factors are independent or cumulative – if the latter is the case, then the advice would only apply when the guarantor is the father and the son is a scholar.

147 bBaba Batra, *ibid.*

148 See Rashba, *Responsa*, part 3, #292.

In these cases, the law considers the intentions of the agent, which cannot be observed, in deciding whether such action is permitted or forbidden. To the outside observer, evasions falling into this category appear to be, or can be interpreted as, permissible actions. Before elaborating on these points, however, I first discuss a case considered relatively simple and uncontroversial: that of one who performs a permissible action which incidentally gives rise to a forbidden outcome.

i. The Immersion of Utensils

Our discussion of evasion began with R. Oshea's ruling permitting evasion of heave offerings and tithes.[149] Here too we begin with one of his rulings allowing evasion, though on another issue.

> R. Hoshea (Oshea) taught: A person may fill an unclean vessel from the well and by evasion immerse it.[150]

On the Sabbath and festivals, the purification of vessels by immersion is prohibited by the Sages. However, if one immerses a vessel in such a way that it does not look as if he wants to immerse it, but as if he wants to draw water from the well, the immersion is permissible. This case highlights two important issues that are central to our discussion: (a) how an action appears to an observer, and (b) evasion in which the action does not reflect the agent's true intentions. Other evasions of a similar nature are also permitted. For example, "We learned: If his bucket or his vessel fell into the well, he may exploit this event as a cleansing immersion."[151] Further, even those who accept R. Judah's view that unintentional actions on the Sabbath are forbidden agree that

> One who wants to rinse the floor on the Sabbath may bring a pail full of water and wash his face from one angle, his hands from another, his feet from yet another, and it happens that his floor becomes rinsed by itself.[152]

The Amoraim comment on this laconic *baraita* that "a wise woman rinses her floor on the Sabbath,"[153] and add "R. Jeremiah and R. Zeira said in the name

---

149 See above, "Changes in Methods of Harvesting."
150 jTerumot 2:1 (2:3 41c); jShabbat, end of ch. 2 (5b–c); jBeitza 2:2 (61b).
151 See the passages cited in note 150.
152 bShabbat 95a.
153 *Ibid.*

of R. Hiya bar Ashi: A clever woman washes glass here, a bowl there, then a pan, and it happens that she rinses her floor on the Sabbath."[154] R. Hiya b. Ashi's position seems to follow from that of his teacher Rab:

> Come and hear: R. Hiya b. Ashi said in the name of Rab: An impure woman who does not have a change of clothing may by evasion immerse herself clothed.[155]

A ritually-impure woman may immerse herself for purification on the Sabbath and festivals, but clothes and vessels must not be rinsed on the Sabbath and festivals. Rab thus advises one who has no clean clothes to immerse herself with her clothes on, thereby evading the prohibition and cleansing her clothes. The Jerusalem Talmud has reservations about the mishnaic prohibition against immersing vessels on the Sabbath and festivals and only prohibits immersion of big vessels. But rather than clearly stating that the immersion of small vessels is permissible, the Jerusalem Talmud, on the basis of permissible evasions in the various *baraitot*, declares: "This Mishnah concerns big vessels but as for small vessels, one may evade and immerse them."[156]

ii. Extracting an Animal from a Pit in order to Slaughter It

One of the most important passages dealing with evasion is the case of "a dam and her calf that fell into a pit," over which the Tannaim R. Eliezer and R. Joshua disagree. The Tosefta relates the case as follows:

> About a dam and her calf that fell into a pit, R. Eliezer says, He brings up the first in order to slaughter it, and slaughters it, and he feeds the other where it is so

154 jTerumot, *loc. cit.*; jShabbat, *loc. cit.* and 16:3.

155 bBeitza 18a.

156 jTerumot, *loc. cit.* and end of ch. 2 (41d); jBeitza, *loc. cit.* This passage is of special interest because it sheds light on the editorial methods of the Jerusalem Talmud. It appears with three different Mishnayot, restricting their scope. The passage begins: "This Mishnah concerns big vessels but as for small vessels, one may evade and immerse them. And so taught R. Hoshea: a man may fill... ." Not one but three Mishnayot are thus restricted. When the passage is quoted in Terumot, it is in the context of a Mishnah which states, "One who immerses vessels on the Sabbath – if inadvertently, he may use them, if intentionally, he may not" (Mishnah Terumot 2:1). In jShabbat the passage is appended to a Mishnah which deals with the eve of the Sabbath at sundown: "it is doubtful whether it is dark or not... and vessels may not be immersed." In jBeitza it relates to a Mishnah regarding a festival that falls on a Sunday: "the vessels must be immersed... before the Sabbath" (Mishnah Beitza 2:2).

> that it does not die. R. Joshua says, He brings up the first one in order to slaughter it but does not slaughter it, but evades and brings up the second. If he does not want to slaughter one of them – this is permissible.[157]

According to biblical law, it is prohibited to slaughter an animal and its offspring on the same day.[158] A dam and her calf fall into a pit on a festival and their owner wishes to extricate them. If he wants to slaughter an animal for the festive meal, he may extricate it. But if he does not intend to do so, the animal is not to be touched until the end of the festival (*muktze*). Since the case in question involves a dam and her calf, which may not be slaughtered on the same day, it follows that at least one of the animals must remain untouched. This is the view of R. Eliezer. That is, only one of the animals may be extricated from the pit by its owner in order to be slaughtered, and it must in fact be slaughtered. R. Joshua suggests a way of evading this outcome: one animal is brought up to be slaughtered, but it can be examined by its owner and judged unsuitable for the festive meal. The owner can then decide he wishes to slaughter the second animal, which can in turn be brought up. Apparently, R. Joshua goes even further. He does not say that the animal has to be killed at once. He permits the owner to say that he has changed his mind and does not feel like eating meat after all and no longer wishes to slaughter either of the animals.[159]

Though not stated explicitly, it is implied that R. Joshua's far-reaching permission is intended to prevent cruelty to animals[160] and does not constitute a general license to take evasive action. In a case where the owner really does not want to eat the meat that very day, if no cruelty is involved, he can wait until the evening, as R. Eliezer suggests. This also follows from other pericopes in the Talmud that refer to this case. The editors of these passages were of the opinion that the reason for R. Joshua's permission to evade was, indeed, to prevent cruelty to animals.

The Gemara compares the case of the dam and her calf to another case, that of the extrication of a firstling which falls into a pit on a festival and may be maimed. The editor of the discussion argues: we could have mistakenly

157 Tosefta Beitza 3:2.

158 Lev. 22:28; chapter 5 of Tractate Hulin.

159 This is implied by a number of variant readings of the Tosefta and by the Jerusalem Talmud. But the Babylonian Talmud apparently holds that one of them must be slaughtered. The commentaries are also divided on the matter. For a summary, see Lieberman, *op. cit.*, Yom Tov, ch. 3, pp. 965–6.

thought that the firstling may be extricated on the basis of R. Joshua's opinion, in order to prevent cruelty to animals, but this is not so.[161] We can infer that an attempt was made to extend the scope of permitted evasions on the basis of "the dam and her calf," but did not succeed.

Similar attempts are reported elsewhere, as in the *baraita* on the case of a barrel that broke on the Sabbath:

> His barrel broke on the roof.... If he happens to have guests he brings another vessel and collects it into the other vessel. He must not first collect it and then invite guests, but he invites first and collects after. And one should not evade in this manner. It was reported in the name of R. Jose bar Judah that evasion is permitted.[162]

The *baraita* does not state exactly what kind of evasion was at issue here. Rashi explains that the Sages were referring to the invitation of guests who do not have to eat.[163] The Gemara tries unsuccessfully to link this argument with the opinions of R. Joshua and R. Eliezer on "the dam and her calf." Usually, the Gemara states, even R. Eliezer agrees that evasion should be permitted, however, in the case of "the dam and her calf," because the animals could have been fed while remaining in the pit, evasion in order to extricate them is not justified. On the other hand, while R. Joshua might ordinarily conclude that evasion should not be permitted, in the case of "the dam and her calf," he permits evasion so that the animals do not suffer. The Gemara concludes that it is impossible to arrive at a general principle regarding evasion from either the opinion of R. Joshua or that of R. Eliezer.[164]

As the example of "the barrel that broke" illustrates, in exploring the question of evasion, the Gemara tries to derive a general principle from the *baraita* of "the dam and her calf" that will apply in other cases. Such attempts appear a number of times in both the Jerusalem Talmud and the Babylonian Talmud; all are unsuccessful. Each time "the dam and her calf" is compared to another case, the analogy fails, though for different reasons. In two instances, we saw, the conclusion is that a general rule cannot be inferred from the opinion

---

160 As to whether an ordinary animal that falls into a pit may be brought up in order to prevent cruelty to animals, see Lieberman, *ibid.*

161 bBeitza 26a; See Y. Gilat, *Mishnato shel R. Eliezer b. Hyrcanos* (Tel Aviv: 1968), p. 210.

162 bShabbat 117b.

163 Rashi *ad loc.* s.v. *ein maarimin bekakh* .

164 bShabbat, *loc. cit.*

of R. Joshua since "the dam and her calf" is a special case, as it involves cruelty to animals. In the last case we considered, the explanation offered for R. Eliezer's position is that "the dam and her calf" does not call for evading since the animals can be fed in any event. Another passage in which a situation involving evasion is compared with "the dam and her calf" mentions only "he feeds the other where it is" as a reason for differentiating between the two cases, and cruelty to animals is not mentioned.[165]

In the Jerusalem Talmud there is a case that seems parallel to "the dam and her calf." In Tractate Pesahim, the Mishnah addresses the problem of dough which becomes impure on the Passover festival before the priestly portion has been set aside.[166] On all other festivals it would be possible to wait till sundown[167] and then resolve the problem, but on Passover this is impossible since the dough might leaven. Given this situation, the Jerusalem Talmud states, R. Eliezer allows evasion, as is related in the Mishnah: "He can evade and say, I want to eat this and this too, and bake it all. When he takes it out of the oven he can say, This I want to keep for later, this I want to keep for later, and leave it for later."[168] It is clear that this procedure is an exigency measure resorted to as a result of the extraordinary circumstances of the case.

R. Joshua says that all this is misguided. In his opinion the prohibition of leaven on Passover does not apply to holy things, and since it is only the priestly portion that is at risk, it is permitted to set aside the priestly portion and then wait until sundown, even if the dough indeed leavens during the day.

The Jerusalem Talmud finds the dispute surprising, since in this case R. Eliezer favors evasion and R. Joshua objects, while in the case of "the dam and her calf" their positions are reversed. Faced with this apparent paradox, the editor of the pericope suggests emending one of the texts by transposing the names of R. Eliezer and R. Joshua. The Gemara does not accept this suggestion and leaves both texts unchanged; however, it brings a number of arguments to differentiate between the two cases.[169]

---

165 bBeitza 37a; bShabbat 154a discusses whether produce may be dropped down a chimney on a festival [to keep it from being rained on].

166 Mishnah Pesahim 3:3.

167 On this Mishnah, see the classical commentaries as well as the *Korban Haeida* and *Pnei Moshe*.

168 jPesahim 3:3 (30a–b); jBeitza 3:5 [3:4] (62a); cf. Gilat, *op. cit.* (note 161), pp. 151–2.

169 In my opinion, the Jerusalem Talmud is imprecise here: it is difficult to interpret R. Joshua's opposition to evasion as a matter of principle. Rather, he has specific

Why does R. Eliezer permit evasion in the case of "the unclean dough" but object to it in that of "the dam and her calf," and why does R. Joshua object to evasion in the case of "the unclean dough" but condone it in that of "the dam and her calf"? R. Eliezer thinks that a solution must be found for the dough problem because unlike that of "the dam and her calf," it involves two strict biblical prohibitions – "it shall not be seen" and "it shall not be found." The Jerusalem Talmud offers two explanations for why R. Joshua permits evasion in "the dam and her calf" but prohibits it in "the unclean dough". R. Edi offers the explanation that the case of "the dam and her calf" involves only transgression of a Rabbinic law, so evasion is permitted, but if we were to adopt the position of R. Eliezer condoning evasion of the strict laws of Passover, this might lead to a more serious transgression. It is not clear which transgression he has in mind,[170] but it is undoubtedly a transgression of biblical law. R. Jose b. R. Bun has a different explanation of the difference between the cases: in the case of "the dam and her calf," R. Joshua permits evasion "to be merciful with the property of Israel" – i.e., so that the second animal will not die or be damaged in the pit. This consideration does not apply in the case of "the unclean dough."

These points, and particularly the controversy between R. Eliezer and R. Joshua regarding "the dam and her calf," make it easier to understand a number of other disputes between the Sages over similar questions.

iii. Salvaging Food from a Fire on the Sabbath

A case similar to that of "the dam and her calf" about a fire on the Sabbath eve is found in the Tosefta and the Babylonian Talmud. By law, in such a case it is permitted to salvage sufficient food for three meals. In order to save more, it is permitted to first gather bread of inferior quality, even several loaves, and then to declare that since there is some better bread, it too should be salvaged and eaten.[171] This conduct is permitted even though it is a premeditated strategy to save more loaves than are really needed for three

---

concerns about "the unclean dough." I think the interesting question has to do with R. Eliezer. Nevertheless, the Jerusalem Talmud cannot be ignored, and it clearly indicates that R. Joshua prohibits evasion.

170 The classical commentators on the Jerusalem Talmud disagree as to this transgression; on jPesahim 3:3, see *Pnei Moshe* s.v. *amar R. Edi* and *Korban Haeida* s.v. *vekan hiyuv hatat*; on jBeitza 3:5, see *Korban Haeida* s.v. *vekan hiyuv hatat*.

171 Tosefta Shabbat 13:7 (14:7), bShabbat 117b.

meals.[172] But open evasion with no pretense at all, such as saving the better bread first, and then taking the bread made of coarser flour, with the excuse that one prefers it, is prohibited. This is the upshot of the discussion in the Babylonian Talmud. At first glance, it seems that the Tosefta can be interpreted to mean that R. Jose b. Judah permits such an evasion. Apparently, the Tosefta has to be interpreted in a such a way that it is compatible with the Babylonian Talmud. The Tosefta states:

> He saved fine bread, he is not allowed to save coarse bread. [He saved] coarse bread, he is allowed to save fine bread. And evasion is not acceptable.

---

172 In the context of a fire on the Sabbath, it is also permitted to invite guests for the meal to save more food from the conflagration. This is not regarded by the Talmud as an act of evasion, even though it is quite likely that the guests would not have been invited were it not for the fire. What the Gemara considers a possible evasion is saving a great quantity of food, which is prohibited by the halakha, and only afterwards inviting guests for a meal. Some authorities permit this evasion, others prohibit it. R. Jose b. Judah permits it (Tosefta Shabbat 13 (14):7, jShabbat 16:3 [15d]). In the Jerusalem Talmud the name given is R. Jose b. R. Bun, but it is clear that the correct reading is R. Jose bar Judah. See note 64 above. A similar case is that of "the broken barrel," mentioned above in "Extracting an Animal... ." In that instance, it was permitted to save more than enough for the host's immediate needs by inviting guests, on the condition that the guests were invited first and the drink salvaged afterwards. This case was compared to the dispute between R. Eliezer and R. Joshua concerning "the dam and her calf." The analogy between the two cases was rejected by the Talmud for various reasons; see bShabbat 117b. However, it is clear that the Talmud regards this kind of action as an evasion. Apparently the evasion of "the broken barrel" is analogous to that of "saving food from the fire," as indicated by the almost identical wording of the pericopes. Indeed, it was so understood by some of the Early Authorities: see, e.g., Maimonides, *Code*, Laws Concerning the Sabbath 25:16; *Magen Avraham* on the *Shulhan Arukh*, Orah Haim 335:3. Nevertheless, a different interpretation of "the broken barrel" has been offered: it is unanimously accepted that the food must not be saved before the invitation; the question is whether it is permissible "to invite guests, who do not have to eat, and would leave the food" (Rashi, Shabbat 117b s.v. *ein maarimin*). Even though the Gemara itself suggests the evasions in these pericopes, since they only involve transgressions of minor Rabbinic prohibitions, namely, enactments to prevent extra vessels from being carried through forbidden places, the temptation to draw sweeping conclusions should be resisted. It is quite possible that the Sages who permit evasion in these cases do so more because they are willing to waive Rabbinic enactments under special circumstances than out of willingness to extend the legal scope of evasion.

The next sentence, "R. Jose b. Judah says evasion is acceptable," apparently refers to a different matter which was discussed directly before, "He saved fine bread."

iv. Preparing Food on a Festival

From a case similar to those we have been discussing, we learn that R. Jose b. Judah is lenient on questions of evasion. It is forbidden to prepare food on a festival unless it will be used that day.[173] Several *baraitot* list examples of preparations which are prohibited unless the person performing them sincerely intends to eat the foods thus prepared. For example: "It is permitted to brew beer on the festival for the needs of the festival, but if not for the needs of the festival it is prohibited."[174] R. Jose b. Judah, however, permits the brewing of new beer even if there is old beer: the brewer may evade and drink from the new beer. That is, in condoning the brewing of beer which is not really needed, since there is enough without it – provided it is consumed on the festival – R. Jose b. Judah, unlike other Sages, permits evasion.[175]

In one of the two pericopes that address this law, R. Jose b. Judah's position is explained as follows. A casual observer would not suspect someone he saw brewing beer of being engaged in prohibited action, since the observer does not know whether the brewer has old beer, and will assume that he does not, and that he is making beer because he truly wishes to drink it on the festival.[176] The wording of the relevant *baraitot* implies that the brewer must indeed drink at least some of the new beer on the festival, so that the beer brewed on the festival is in fact consumed on the festival. This discussion, then, clearly invokes the principle of "evasions not immediately recognizable as such," as Maimonides puts it.[177] This principle is an important, and perhaps even crucial, factor in deciding whether an evasion is permitted.

A very similar form of evasion is found in Tractate Beitza. Someone cooks many pots of food and says they are all for the festival, or, after having cooked, says, on the festival, that the food is for the next day, and cooks again for the festival. If it is clear that not all the pots are for the festival, but rather, some are for the following day, he must not eat from the food in the pots

173 Pesahim 46b, 62a; bBeitza 21a.
174 bShabbat 139b, bMoed Katan 12b.
175 bMoed Katan, *ibid.*
176 bShabbat, *ibid.*
177 Maimonides, *Code*, Laws Concerning Repose on a Festival 7:8.

which were left.[178] Though the Talmud calls this an evasion, it is very close to outright deception. The innovation in this ruling is that the Sages treat this evasion as worse than a demonstrative, wilful action, such as saying deliberately that one is cooking for both the festival and the following day.[179]

v. Salvaging Clothes from a Fire on the Sabbath

An example of evasion similar to those just discussed is found in the Mishnah; it is doubtful, however, if it should be classified as an evasion, especially since the act itself is not forbidden. It too concerns a fire on the Sabbath. The Mishnah relates:

> And he puts on as many clothes as he can and he cloaks himself with as many cloaks as he can. R. Jose says: eighteen garments and he goes back and puts on and takes them out. And he tells the others, come and save them with me.[180]

---

178 bBeitza 17b Cf. jBeitza 2:1 (61b). Nevertheless, other acts, close to evasion, were permitted:

> The Sages taught: One may not bake on a festival for the needs of another festival. But they also said, a woman may fill the pot with meat though she needs only one piece, the baker may fill the barrel with water though he needs only one pail; but he may only bake as much as he needs. R. Simon b. Eleazar says: A woman may fill the oven with loaves because bread baked in a full oven is better. Rava says: The law is according to R. Simon b. Eleazar (bBeitza 17b).

179 Commentators on the Talmud have been much preoccupied with this pericope because of its uncompromising attitude toward evasion; it appears to contradict others in which evasion is permitted. See, e.g., *Tosafot* on Beitza 17b s.v. *ubilvad; Hidushei Harashba* on Beitza 11b s.v. *Rav Ada b. Ahava; Hamikhtam* on Beitza 11b, 17b; *Peirush Talmid Haramban* on Beitza 17b s.v. *veim*. But as we said, there is no particular difficulty here, since the pericope deals with an action which borders on deception, though the term "evasion" is used. The verb *maarim* is used by the Talmud to denote both evasion and deception. In this case, the evasion is tantamount to intentional violation of the law. In both cases, someone knowingly cooks on a festival for the day after. When it is a case of intentional violation, he admits this, while if it is evasion he says that some is left over. Apparently, this is a matter which is patently obvious: a transparent lie. Still, R. Ashi, who makes the statement, seems to object to leniency in matters of evasion. See below, note 192.

180 Mishnah Shabbat 16:4.

A *baraita* recounts the case as follows:

> He puts on garments and goes out and takes them off and he enters again, and puts them on and goes out and takes them off even all day long, says R. Meir. R. Jose says: eighteen garments.[181]

That is, according to R. Meir, one may dress and undress all day long in order to salvage clothes from a fire. Since dressing is not prohibited on the Sabbath, it is clear why R. Meir suggests this action, the purpose of which is clearly to salvage as many clothes as possible from the fire. Rather, it is the opinion of R. Jose which requires explanation, but that is not our concern here, since his reservations apparently have nothing to do with evasion.

vi. Changing One's Mind on Festivals

Another example of evasion via intent, involving someone who knows in advance that he is going to change his mind, is mentioned in a *baraita*. Indeed, it is a classic example of this type of evasion. Proposed by R. Judah, and opposed by R. Meir and R. Jose for reasons not germane to the question of evasion, it concerns the writing of scrolls for *tefilin* and *mezuzot* on the intermediate days of festivals (*hol hamoed*). R. Judah and R. Meir both contend that on the intermediate days it is forbidden to write these scrolls for sale, though permitted for personal use. However, "R. Judah says: He may evade and sell his own and then write new ones for himself."[182] That is, according to R. Judah, he may write the scrolls as if for his personal use and then decide that he does not like them, sell them, and write new ones for himself, repeating this sequence a number of times. In fact, as the Gemara tells us, the law follows neither of these views but rather that of R. Jose, who argues that no debate is necessary: "It is permitted to write them and sell them in the usual way in order to make a living."[183] It may be reasonable to infer from this pericope that the ruling on salvaging clothes from a fire on the Sabbath, namely, R. Meir's suggestion (see above), indicates that this is not really a case of evasion, because if it were, there would be a contradiction between R. Meir's position regarding the scrolls, where he disagrees with R. Judah and does not permit the sale, and his position on salvaging the

181 bShabbat 120a.

182 bMoed Katan 19a. It seems the phrase "and he spins on his lap the blue wool for his fringes" was inserted erroneously. See *Dikdukei Soferim*.

183 bMoed Katan, *ibid.*

clothes. However, some of the Amoraim apparently did regard the strategy suggested for salvaging clothes as an evasion.[184]

What was the general attitude of the Amoraim to evasion on festivals? With regard to salting meat [to render it kosher] the ruling is that one may not salt more pieces of meat than are needed for the day. However, Mar Samuel permits salting a number of pieces together. The underlying principle is common to a number of laws concerning actions which have an affect on both items needed for the festival and others which are not needed. The salient ruling for our discussion is that of R. Ada b. Ahava. "He evades and salts it piece after piece."[185] Rashi explains "After he salts the piece needed for that day, he says that he prefers the other piece and salts it too."[186] This is an explicit example of the type of evasion exemplified by "the dam and her calf" and other cases we examined above. The Early Authorities make it clear, here and elsewhere, that the evasion is allowed only for the purpose of enhancing enjoyment of the festival. This purpose is particularly stressed by the classical commentators on the Talmud to justify the Talmud's stance that such evasion is valid and useful.[187]

vii. Carrying Nuts on the Sabbath

On the Sabbath, a woman is permitted to carry a nut attached to her garment because it is regarded as an ornament. Abbaye poses the question: what

---

184 See the following two sections.

185 bBeitza 11b.

186 Rashi *ad loc.* s.v. *maarim.*

187 See: *Hidushei Harashba* on Beitza 11b s.v. *Rav Ada*; "If it were not allowed there is a great likelihood of lessening the joy of the festival (*Hamikhtam*, Beitza 11b); "but because we fear that it may disturb the joy of the festival, this evasion is nevertheless allowed if there is a great need, e.g., salting it piece after piece ... for all these evasions are needed and allowed" (*ibid.*, 17b); *Peirush Talmid Haramban* on Beitza 17b; "Due to the joy of the festival they permitted the means for the end, because if this were not allowed he would not kill the animal and this would lessen the joy of the festival ... and the evasion is permitted for the joy of the festival" (Rashba, *Responsa*, part 1, #681); "It appears there is no disagreement on this evasion because it is for the food needed on the festival, and anyone who might see it thinks that it is for the rejoicing of the festival, but otherwise there is unanimous agreement that evasion is not permitted" (*Hidushei Haritba*, Moed Katan 12b); "Since they permitted evasion in salting meat in order not to disturb the joy of the festival ... if it were not permitted, the joy of the festival would be disturbed ... and many similar things were permitted for the joy of the festival" (*Tashbetz*, part 2, 10).

if she evades and carries on her dress a nut that is purportedly an ornament, but her intention is to feed it to her child? The question is compared to that raised by the "salvaging clothes from a fire" dispute, where it is permitted to put on more clothes than needed or customary. The editor of the pericope contends that as far as permitting or prohibiting evasion is concerned, the two cases are not analogous.

> Abbaye asked: What about a woman who evades and clasps a nut to her dress in order to take it to her young child on the Sabbath? This question is addressed to both those who permit evasion and those who do not. To those who permit evasion in the case of the fire: there it is permitted because if saving clothes by evasion were not allowed, he would simply extinguish the fire, but here, if we do not allow her to carry the nut, she will not carry it. On the other hand, perhaps those who do not permit evasion in the case of the fire so rule because the clothes are carried in the usual manner, but here the nut is not carried in the usual manner and therefore it may be permitted. Unresolved.[188]

Apparently, the prohibition stems from the agent's intention: whether the woman intended the nut to be part of her apparel, or sought a way to give it to the child later on. No explicit conflict between the Sages over this passage is recorded, and no decision is reached.[189]

viii. Subjective Intention

We now examine two additional examples of evasion in which the decisive factors are appearances and the agent's subjective intent.

The first case concerns the use of a yeast strainer on a festival. This is forbidden, because using the strainer is a workaday act – ordinarily, the strainer is used for straining yeast. But if one wishes to put pomegranates into the strainer, handling it is permitted. Since, as we said, the strainer is usually used for preparing yeast, the law states that if, on a festival, pomegranates have been placed in the strainer, which is permitted, it is then permissible to add yeast. At the outset the strainer is put up as if for the purpose of storing fruit, but the person who hangs it knows that his real intention is to prepare yeast. Putting a few pomegranates in the strainer first makes it legally permissible to add yeast subsequently.

188 bShabbat 65b.

189 I doubt whether salvaging clothes on the Sabbath, which has been described as an evasion, would be so classified on our definition of evasion.

The Gemara stresses its concern with appearances after comparing this ruling to that of "brewing beer on a festival." As noted, it is permitted to brew new beer even though old beer is available. The Gemara explains that in the case of brewing beer there is no need for subterfuge – such as putting a few pomegranates in the strainer – because anyone who sees beer being brewed will assume the brewer has no beer; but in the case of the strainer, were it not for the pomegranates, it would be clear that it was hung there to prepare yeast on the festival, which is forbidden.[190]

Another case in which the subjective intention of the agent is emphasized is that of the law forbidding the plugging of holes on the Sabbath. The Talmud tells of a scholar who plugged a hole on the Sabbath with a piece of food, claiming that he merely wanted to store the food in a safe place. R. Ashi points out, however, that the anecdote relates to a learned man, and we cannot conclude that such conduct is universally permissible.[191] In my opinion, R. Ashi's reservations reveal an atypical attitude to evasion which arose at the end of the Amoraic period and do not reflect the accepted law of the Tannaitic period.[192]

---

190 bShabbat 139b; see above near note 174ff.

191 *Ibid.*

192 In my opinion, R. Ashi tried to limit the scope of permissible evasion. In this pericope R. Ashi justifies evasion in various ways. He is the first – and only – of the Sages to condone an act of evasion because the law it circumvents is a Rabbinic prohibition. This opinion caused much debate among the Early and Later Authorities. R. Ashi is also the one who condoned an evasion because the evader was "a learned man," implying that such behavior is unacceptable when practiced by ordinary people. R. Ashi, I believe, is consistent in his attempts to limit evasion that had been going on for generations. As we noted, he states that the Sages regard evasion more harshly than intentional transgression (see above, "Preparing Food on a Festival"). R. Ashi is the one who restricts a statement apparently made by Rabba bar R. Huna. On the face of it, Rabba bar R. Huna seems to have said, regarding the "making yeast on a festival" case, that one can change one's mind and put in yeast rather than fruit after hanging up a strainer, even without having put in any pomegranates. However, it must be kept in mind that all the instances of the second type of evasion involve Rabbinic prohibitions. It is quite possible that this type of evasion – bordering on outright deception but permitted on the advice of the authorities themselves – was allowed if the law evaded was a Rabbinic prohibition. Whether evasion of a biblical prohibition is similarly permitted has been considered at length both by the halakhic authorities and by contemporary scholars. However, none have distinguished between the different kinds of evasion, weakening the argument of those who

The evasion here is similar to cases already considered. Someone puts a chunk of food in a crack and claims that he does so for a purpose which is permissible – to store food – though it is known that his intention is actually to perform a forbidden action – mending the crack. In other words, an act may be permitted if performed with one intention, but forbidden if performed with another. Since no observer can know the intention prompting the act, it is always possible to claim that it was performed with permissible intentions. As a rule, however, this is not done, since it would trivialize the halakha. It is acceptable, though, if the Sages regard the overall purpose of the evasive act as desirable.

Our examination indicates, then, that acts falling under the second type of evasion are permitted or forbidden on the basis of the following criterion: if the general aim of the action is positive, the halakha permits evasion, provided that its appearance is such that it can be interpreted by others as a permissible act. The intentions of the evader are not scrutinized. Apparently, when the goal of the evasion was desirable, the Sages wanted to enable the agent to achieve it, but not at the cost of rescinding the law that ordinarily applied to the case. They thus prefer evasion by means of actions that appear compatible with the law to amending the law itself.

B. Laws of Purity

Thus far, we have considered disputes over the permissibility of evasive actions. We will now consider prohibited evasions. Although the acts of evasion we will discuss seem similar to those already examined, they are deemed impermissible because they do not satisfy the criterion formulated above: they lack a positive overall aim that compensates for the inherently negative aspects of evasion. In my opinion, by explaining why the evasion is not valid in these cases, although in similar cases it is accepted as legitimate,

---

limit the permissibility of evasion to Rabbinic prohibitions. If, however, we make the suggested distinction between the two types of evasion, it becomes clear that in cases of the first type, evasion is valid even for biblical prohibitions, while with regard to the second type, it seems possible that evasion is indeed restricted to Rabbinic prohibitions. For a discussion of evasion of Rabbinic and biblical prohibitions, see: B. Benas, "The Legal Device in Jewish Law," *Journal of Comparative Legislation and International Law* (1929), 3rd series; vol. 2, p. 75, and E. Berkowitz, *Hahalakha, Koha Vetafkida* (Jerusalem: 1981), pp. 118ff.

formulation of this criterion resolves the cardinal problem confronting researchers in this area.

The Tosefta brings a number of laws that specify how an act that usually renders a substance ritually pure can bring about the opposite effect if the person performing it intends evasion. The following expression is appended in each case: "if he evaded, they are impure." The laws in question cover a variety of matters related to the laws of purity and impurity, but have one thing in common: an act performed with a deceitful intention brings about impurity, though had the same act been performed sincerely, no impurity would have ensued.

The only *baraita* on this subject that gives names of Tannaim deals with olives that were soaked in water and congealed into a clump. The Tosefta says that as long as the soaking, or preparation, process is still going on, it is possible to take some of the olives to press without rendering the entire clump impure, provided the clump itself remains in place. If the clump is removed, however, the olives are impure. The *baraita* then mentions the opinion of R. Ishmael b. R. Jose, who says in the name of his father: "Even if he extracted the clump from its place, they are pure, as long as he does not evade; if he evades, they are impure."[193] That is, if someone intended to remove some of the olives and the whole clump came out accidentally, although technically this indicates that the preparation stage is over, which implies impurity, both the olives and the clump are still pure. However, if his real intention is to remove the entire clump, but he tries to conceal this, pretending he is taking only some of the olives, then "accidentally" removing them all, the entire clump is impure. In other words, taking the clump by accident does not impact on its purity, but pretense that an accident has occurred renders it impure.

When an unlearned person brings to a learned one a vessel intended for an offering, if it is for a sin offering, the presumption is that he brings a pure vessel. If the vessel is intended for the priestly heave offering, however, there is no such presumption, and the vessel may not be used. If the learned man orders a vessel for the sin offering and the unlearned one brings one for the sin offering and another for the heave offering, both may be used. The law apparently reasons that the ignorant man takes greater care when a sin offering is involved because of its importance, but tends to be negligent if

193 Tosefta Toharot 10:12; cf. Mishnah Toharot 9:7. The Mishnah does not mention evasion; see R. Samson of Sens' commentary on the Mishnah *ad loc.*

the purity required is only that specified for heave offerings. If he brings two vessels, he is careful with both. But if the learned man needs a vessel for the heave offering and to ensure its purity tells the unlearned man that he needs it for a sin offering, the Tosefta says that the vessel is impure. The Sages thus penalize the learned man in this type of case, since it can be assumed that, thinking it is for a sin offering, the ignorant man indeed takes proper care of the vessel, and that it is, in fact, pure.[194] The reason for this ruling is not mentioned in the Tosefta but it is tempting to speculate. Perhaps the lie told by the learned man is sufficient reason for his punishment by the Sages. It may well be that had such an evasion been accepted, it would have become the standard procedure, and unlearned men, aware that requests made by learned men for vessels to be used for sin offerings need not be taken at face value, would cease to ensure the purity of vessels they provided.

Here too, identical requests made by the same individual bring about different outcomes depending on his true intentions: if he really intends to use the pot for a sin offering, he may even use it for a heave offering; if he does not really intend to use it for a sin offering, the vessel is impure. Here, as elsewhere, the Tosefta adds: "as long as he does not evade; if he evades, they are impure."

A similar law governs the purity of basketwork:

> All untrimmed woven baskets in temporary use are unclean. If they were intended to be brimmed and trimmed, though they are casually used in their present state and thrown away, they are pure, as long as he does not evade; if he evades, they are impure.[195]

If the intention was truly to give the baskets a professional finish, to tie the brims and trim the protruding ends, then their temporary use does not render them impure. But if the weaver does not intend to give the baskets these final touches, then the same casual usage causes impurity. He cannot evade and temporarily use them under the pretense that he will finish them later.

---

194 Tosefta Hagiga 3:27; Tosefta Para 4:13; S. Lieberman, *Tosefet Rishonim*, Hagiga *ad loc*. Cf. Mishnah Para 8:1 and R. Samson of Sens' commentary. My interpretation of the Mishnah follows R. Elijah of Vilna and S. Lieberman rather than the Rash.

195 Tosefta Keilim 5:14; cf. Mishnah Keilim 16:3 and R. Samson of Sens' commentary; also Maimonides, *Code*, Laws Concerning Utensils 6:14.

There is yet another law in which the end of the production process is the criterion for whether impurity has been incurred:

> He has a clump of olives and wants to keep it pure, though he intends to add a measure or two more – they are pure, as long as he does not evade; if he evades, they are impure.[196]

That is, it is forbidden to intentionally leave some olives outside the receptacle, to be added later, in order to artificially prolong the production stage (recall that the olives become impure once it is over).

Finally, consider an example from the laws governing the impurity of a corpse. It is based on the rule that the doorway through which the corpse will be removed immediately contracts impurity:

> The corpse is in a house with many exits… . He intended to remove it through one of them, the others are pure. There was an opening to the north and one to the south, and he planned to take it through the north, but later his brothers and relatives came and said, "It shall not be taken out but through the south." The south renders the north pure, as long as he does not evade; if he evades, they [both entrances] are impure.[197]

These examples indicate an unequivocally negative attitude to evasion, which is not surprising. As we have seen, when the acceptability of evasion turns on subjective intention, evasion is only condoned if it serves a purpose that is basically desirable and justifies tacit acquiescence, and, if the act itself can be interpreted by an observer as a permissible one. Only if both these conditions are met are the Sages willing to recognize an evasion as valid. As I said, I believe this test can help resolve instances of contradictory rulings in what appear to be similar cases.

However, it is also possible that the Sages' negative attitude to evasion in matters related to the laws of purity can be explained differently. In this

---

196 Tosefta Toharot 10:5. Cf. Mishnah Toharot 9:3 and R. Samson of Sens' commentary; Maimonides, *Code*, Laws Concerning The Uncleanness of Foodstuffs 11:8–10. The main point is in the Mishnah, but here too, the appended remark regarding evasion is missing.

197 Tosefta Ohalot 8:4. Cf. Mishnah Ohalot 7:3, in which the same rule appears, again, without the appended remark about evasion, and Rash's commentary; Maimonides, *Code*, Laws Concerning Corpse Uncleanness 7:2; cf. bBeitza 10a and the novellae of *Shita Mekubetzet ad loc.*

realm, decisions and intentions often play an important role, and not infrequently, an object's status as regards purity or impurity varies according to the intentions of its user. It is reasonable to suppose, therefore, that the salient criterion for evaluating evasion in matters of purity is not whether the act's aim is desirable or not, but rather, the special nature of the laws of purity themselves.

4. Conclusion

The main points which have emerged from this study of evasion are as follows:

1. Evasion is recommended if the ultimate aim of the evasive act is commendable in the opinion of the legal authorities. The Sages overcame their distaste for the act of evasion by focusing on the anticipated positive outcome. We examined a number of examples of evasion undertaken for a praiseworthy cause, such as arranging maintenance for a needy scholar and his family, preventing cruelty to animals, R. Tarfon's act of feeding hungry women, and the general principle that assets of Jews should be spared excessive taxation. It appears that at least some of the controversies among the Sages regarding cases of what we called the second type of evasion can be explained by the fact that in these cases there is some doubt as to whether the goal motivating the evasion is worthy enough to justify evasion.

   The idea of a law's being set aside for the sake of an outcome that appears more just is not foreign to Jewish law. For example, the basis for the policy calling for behavior "above and beyond the letter of the law" is the maxim that the destruction of Jerusalem came about "because they applied the Torah strictly."[198]
2. Evasion is prohibited if the act's ultimate purpose is not desirable. Nevertheless, evasions falling under the first of the two categories we defined are valid, though forbidden.[199] As to the second category, here, the intention is an integral part of the act. If the intention is not desirable, the act is invalid. In other words, in evasions of the first type, such as sale

198 bBaba Metzia 30b; see S. Shilo, "On One Aspect of Law and Morals in Jewish Law: Lifnim Mishurat Hadin," 13 *Israel Law Review* (1978), p. 359.

199 Here, as in other areas of Jewish law, though the act itself is negative and forbidden, it is nonetheless valid. This thus constitutes an example of a wider issue in Jewish law, the many cases defined as "forbidden but valid" (*lex imperfecta*), a question beyond the scope of the present discussion.

to a third party, the act itself is perfectly sound from a legal point of view, and only the motive is improper, hence the evasion itself is valid. But with regard to evasions of the second type, such as those having to do with observance of the Sabbath, festivals and laws of purity, the intention of the agent is of crucial importance in determining whether a given act is valid or null and void, permitted or prohibited. The failure of attempted evasion in these circumstances can be attributed to unseemly intentions.

3. Evasions must preserve, at least externally, the appearance of legality. The Sages apparently felt this was a reasonable price for an end they considered desirable. The formal aspects of the law are thus of paramount importance here; in this, the halakha resembles other ancient legal systems.
4. Appearances strongly influence the Sages' attitude to various kinds of evasion, especially those of the second type. Explicitly or implicitly, a number of sources relate that one of the points that concerned the Sages when discussing a case of evasion was whether an observer of the act in question could conclude that it was legal. In talmudic terminology this question is phrased, "Is the thing obvious?", while in his *Code*, Maimonides says of a certain evasion: "This evasion is not recognized as such by an observer."[200] The point is discussed directly in only one passage, but it is of great significance, particularly as concerns evasions of the second type, and its centrality is implicit in the other pericopes as well.
5. Invoking one law to evade the consequences of another is always valid, even if the action itself is unseemly. The means of preventing such circumvention, however, lie beyond the laws of evasion. Legislation may be enacted to invalidate the outcome of the evasion (e.g., abandonment of property only comes into effect after three days), or personal fines may be imposed on the basis of the maxim, "As he has done so shall be done unto him, his due returns upon his head" (from bGitin 40a–b).[201]

---

200 Maimonides, *Code*, Laws Concerning Repose on a Festival 7:8.

201 A similar attitude is found with regard to "an act originating from a transgression." In his comprehensive study of the subject, Shochetman argues that

> Any legal action based on the illegal contract is valid ... the liability itself is valid, though it may not be enforced ... . If the court is of the opinion that justice is not served by enforcing a liability originating from the illegal contract, it may punish the litigant by denying this right [E. Shochetman, *Maase Haba Beaveira* (Jerusalem: 1981), p. 255].

6. The Sages reveal great pragmatism. In permitting an evasion, they considered serious transgressions that would likely be committed were the evasion not condoned.
7. In spite of the numerous passages in the Talmud that discuss evasion directly, the Sages do not compare evasions in different legal spheres. The Sages apparently regard questions of evasion as context-specific and seek *ad hoc* solutions for the various cases, but do not view them as different expressions of a basic principle. They seem to have left this task to us.
8. Most, if not all the cases of evasion we considered, have to do with circumventing an individual's liabilities to the legislator – to the Almighty – and not with evasion of duties between individuals. The halakha has no qualms about declaring illegal instances of the latter, such as defrauding a creditor by fictitious transfer of ownership, though it readily recognizes the fictitious sale of leaven to avoid transgression of the prohibition of leaven on Passover.[202]
9. As is the case regarding any other subject, an understanding of evasion in the Talmud calls for examination of the extent to which external factors played a role in the development of the halakha. We noted various such influences in a number of cases: in the laws pertaining to heave offerings, tithes, firstlings, and tithes on the herd.
10. We should not ignore the frequently-encountered scholarly argument that evasion is a mechanism for circumventing intolerable legal impasses resulting from changing times, for which the halakha could find no other solution. I believe the validity of this argument has been overestimated.[203]
11. It may well be that evasions of the second type are permitted only to elude Rabbinic prohibitions.

Though at the beginning of the paper I made a distinction between evasion of the law and legal fiction, in terms of their objectives they have something in common. What Blackstone said of the objectives of legal fiction is equally true of evasion:

> No fiction shall extend to work an injury; its proper operation being to prevent a mischief, or remedy an inconvenience, that might result from the general rule of law. So true it is, that *in fictione juris semper subsistit aequitas*.[204]

---

202 M. Elon, *Herut Haprat Bedarkei Geviat Hov Bamishpat Haivri* (Jerusalem: 1964); see references to *havrahot vehishtamtuyot hayavim* in the index, p. 298.

203 See, e.g., H. Tchernowitz-Rav Zair, *Toldot Hahalakha* (New York: 1948–50), 2nd ed., vol. 1, pp. 179ff.; and Z. Falk, *Dinei Yisrael Bimei Bayit Sheini* (Tel Aviv: 1969), vol. 1, p. 26.

204 *Blackstone's Commentaries on the Laws of England*, vol. 3. ch. 4, p. 43.

CHAPTER 7

INADVERTENT ADULTERY (*SHGAGA*) IN JEWISH LAW: MISTAKE OF LAW AND MISTAKE OF FACT

Moshe Drori

1. Introduction

The concepts of mistake of law and mistake of fact are central to any legal system, which must define its position vis-à-vis both the individual who acts in good faith, assuming a state of affairs to exist which differs from the true situation (mistake of fact), and the person who acts without knowing that his action is forbidden or illegal (mistake of law).

In Jewish law, the term for inadvertent error is *shgaga*. This article will investigate the place of inadvertent error in laws governing adultery. The conclusions drawn will be used to demarcate the parameters of inadvertent error in other areas of Jewish law, as compared to the role played by mistake of fact and law in the English and Israeli legal systems.

According to Jewish law, the adulterous woman, or, in halakhic terminology, the woman who fornicates with another man[1] while married,[2] is forbidden to remain married to her husband[3] or to marry her lover.[4,5] However, not

---

1 This means actual sexual intercourse, as opposed to promiscuous behavior or remaining alone with a man (*yihud*). See B. Schereschewsky, *Family Law in Israel* (Hebrew), 4th ed., (Jerusalem: 1993), p. 321, n. 75. On promiscuous behavior and *yihud*, see *ibid*., pp. 317–318. And see his "Divorce," in M. Elon, ed., *The Principles of Jewish Law* (Jerusalem: 1975), pp. 414, 418 (henceforth, *Principles*).

2 The laws governing what is termed "fornicating with another man" are applicable only when the actions occurred while she was married. If a woman had intercourse wlth another man prior to her marriage, her husband can, in certain cases, demand a divorce for other reasons, such as: the transaction was based on error, she has become disgusting to me, she is blemished. In special cases the husband may even be able to claim that fraud occurred. See: Appeal 154/1959, husband A.v. wife B., Rabbinical Court Judgments (henceforth RCJ), vol. 3, p. 255; File 4831/1968, husband A.v. wife B., RCJ, vol. 7, p. 272.

3 Mishnah Sota 5:1 (bSota 27b); bKetubot 9a; Rashi's commentary on bSota 18b s.v. *asura liyevama*; *Shulhan Arukh*, Even Haezer 11:1 and *Beit Shmuel ad loc*., par. 1. Cf. Schereschewsky, *op. cit*., pp. 321ff.

4 This law is derived from homiletic exegesis of the biblical account of the wayward woman. See Mishnah Sota 5:1 and a discussion of the exegesis in Albeck's *Commentary on the Mishnah*, Seder Nashim (Jerusalem-Tel Aviv: 1959), p. 384 of the addenda. Even if the adulterous woman marries her partner in adultery, they must divorce. See Mishnah Jebamot 2:8 (bJebamot 24b). For summaries of the discussion in the halakhic literature, see Maimonides, *Code*, Laws Concerning the Wayward Woman 2:12; Tur, Even Haezer 11:1, 178:32; *Shulhan Arukh*, Even Haezer 11:1, 178:17–18. Cf. Shereschewsky, *Family Law*, cited in note 1, p. 322, n. 85, and p. 50, where he discusses the reasons for the prohibition. The prohibition against the adulterous woman's marrying her lover is found in other legal systems, e.g., in German law. Also see the 2nd ed. of *Family Law*, (Jerusalem: 1967), p. 57, n. 38.

5 The adulterous act also has three additional consequences, which will not be discussed here:

1. The adulteress loses her rights to her basic and supplemental maintenance/alimony upon divorce (*ketuba*). See bKetubot 101 a–b; Maimonides, *Code*, Laws Concerning Marriage 24:6 and 10; Tur, Even Haezer 115:19; *Shulhan Arukh* Even Haezer 115:5. This was also the ruling reached by a rabbinical court in Israel in a contemporary case, see: File 300/1956, A.V.B., RCJ, vol. 2, pp. 125–128. Cf. Schereschewsky, *op. cit*., pp. 324–325.
2. The offspring of the adulterous woman are *mamzerim* and cannot join the community of Israel [i.e., they are forbidden to marry Jews]. See Deut. 23:3, Sifre Deut., Ki Tetze, par. 248, Finkelstein-Horowitz edition, pp. 276–277;

every act of adultery makes a woman forbidden to both her husband and her lover.˙Jewish law recognizes two cases in which the laws pertaining to the adulterous woman do not apply, even though a physical act of intercourse occurred: adultery by rape, and inadvertent adultery.[6]

Because of the connection between adultery by coercion and inadvertent adultery, I will first examine the laws governing a married woman who is raped, then devote the greater part of the discussion to inadvertent adultery. Before turning to adultery, however, it will be helpful to briefly survey the extent to which inadvertent error plays a role in Jewish law in general.

2. Inadvertent Error in Jewish Law

The classic example of inadvertent error in the halakhic literature is that of the unwitting performance of a prohibited act on the Sabbath. The law recognizes two kinds of error regarding prohibited activities on the Sabbath: either the transgressor was unaware that the day of the week was the Sabbath, or he did not know that the action in question was forbidden on the Sabbath.[7, 8] This

Mishnah Jebamot 4:13 (bJebamot 49a); Maimonides, *Code*, Laws Concerning Forbidden Intercourse 15:1; Tur, Even Haezer 4:15 and 23; *Shulhan Arukh*, Even Haezer 4:13–15. Cf. Schereschewsky, *op. cit.*, pp. 353–355; and his article on "*Mamzer*," in *Principles* (see note 1), pp. 435–436.

3. On the criminal aspects of the matter, see H.H. Cohn, "Adultery," in *Principles*, pp. 488–489.

6 There is another category of cases in which the question of whether the laws pertaining to the adulterous woman apply is problematic: cases in which a woman remarries after a court mistakenly rules that her first marriage is annulled. This type of case is sometimes thought to be analogous to rape (*ones*). In any event, it raises a number of complex issues related to the fallibility of courts, in particular, that of the common court (*beit din stam*) and the distinguished court (*beit din muvhak*); issues beyond the parameters of this discussion. On this problem, see R. Solomon b. Adret, *Responsa*, part 1, #10 and #1189; *Hut Meshulash* #13, both of which will be discussed below, especially near note 79ff.

7 Mishnah Shabbat 7:1; bShabbat 67bff.; Maimonides, *Code*, Laws Concerning Offerings for Transgressions Committed through Error 2:3.

8 There is a difference of opinion in the Babylonian Talmud *ad loc.*, as to what constitutes knowing that the activity is prohibited on the Sabbath. Does the transgressor only have to know that the punishment for deliberately violating the Sabbath is divine extirpation (*karet*), or must he also know that one who inadvertently violates the Sabbath has to bring a sacrificial offering to the

also applies to the prohibition against eating prohibited fat (*heleb*): one who is unaware that eating prohibited fat is forbidden, or does not realize that what he is eating is prohibited fat, is in both cases considered an inadvertent transgressor.[9]

To use the terminology of modern criminal law, inadvertent transgression is defined as both mistake of fact (the individual does not know that today is the Sabbath or that what he is eating is prohibited fat) and lack of knowledge of the law or mistake of law (the individual does not know that the activity is prohibited on the Sabbath,[10] or that it is forbidden to eat prohibited fat).[11,12] It should be emphasized that while modern criminal law recognizes mistake of fact as an acceptable defense,[13] it does not consider ignorance of the law, or

---

Temple? Maimonides rules that if the transgressor knew that the activity was prohibited, but did not know that one who transgresses the prohibition incurs divine extirpation, he has transgressed inadvertently; if he knew that the transgressor incurs divine extirpation but did not know that the unwitting transgressor has to bring a sacrifice, he is considered to have transgressed intentionally. See Maimonides, *Code*, *ibid.*, 2:2.

9 *Ibid.*, 7:2.

10 See note 8 for details.

11 A similar definition couched in slightly different terminology is found in the *Mishne Lamelekh*, a work by R. Judah Rosanes, who lived in Turkey in the second half of the seventeenth and early eighteenth century. Commenting on Maimonides' *Code*, *loc. cit.*, he writes:

> A general principle to be derived from the discussion at bShabbat 67b is that there are two types of inadvertent transgression: one concerns the facts (*bimetziut*), for instance, a person ate what he thought to be permitted fat and it was *heleb*; the second error concerns the law (*badin*), for example: he thought that a certain activity was permissible or did not entail divine extirpation, and later found out that it was forbidden.

12 The role of *shgaga* in the laws of manslaughter differs from that which it plays in the laws of forbidden fat and the Sabbath; for more on this, see below.

13 See, for example, article 17 of the Penal Law, 5737–1977, formerly article 12 of the Mandatory Criminal Code Ordinance, 1936:

> Unless otherwise provided either expressly or by implication, a person who does or omits to do an act upon an honest and reasonable, but mistaken, assumption as to a state of things, is not criminally responsible to any greater extent than he would be if the state of things were such as he assumed it to be.

And see the detailed discussion below.

mistake of law, a legitimate excuse: *Ignorance of the law is no excuse*.[14] In Jewish law, however, the concept of inadvertent error encompasses both mistake of law and mistake of fact.

3. Adultery under Duress (*ones*)

Because of the close connection between unwitting adultery and adultery committed under duress, the laws regarding a married woman who was raped will be briefly reviewed.

The laws of the adulterous woman do not apply to a woman who was raped, that is, one who was forced to have sexual intercourse against her free will. She is permitted to remain with her husband. This law is derived from the biblical verse concerning the wayward woman (*sota*):

> And a man lies with her carnally, and it be hid from the eyes of her husband, and she being defiled secretly, and there be no witness against her, neither she be taken in the act.[15]

On the phrase "neither she be taken," Sifre comments, "this excludes the woman who was raped."[16] A *baraita* in the Babylonian Talmud further explains: "neither she be taken – she is forbidden; if she were taken – she is permitted."[17, 18] The *baraita* understands the words "neither she" to refer to a particular case, that of a woman who is not taken, but is wayward. Hence in the case of another, who is seized and raped, she is not considered wayward, and

14 This maxim is the basis for article 12, Penal Law 5737–1977, formerly article 8 of the Mandatory Criminal Code, 1936: "Ignorance of the law shall not serve as a ground for exemption from responsibility for an offence unless knowledge of the law is expressly declared to be an element of the offence." This will be discussed at length below.

15 Num. 5:13.

16 See Sifre on Num., Naso, par. 7 (Horowitz ed., p. 12).

17 bKetubot 51b.

18 The text quoted above is at the beginning of the *baraita*. The Babylonian Talmud cites two traditions as to the end of the *baraita*: one concerning a woman whose adultery began under duress and then became voluntary, who is permitted to her husband, and another concerning the wife of a priest (*kohen*), who is forbidden to her husband even though she was raped, i.e., her adultery is coerced rather than voluntary.

is therefore permitted to remain with her husband.[19, 20] Maimonides summarizes the laws concerning the woman who was raped as follows:

> If a married woman has fornicated with another man inadvertently (*beshogeg*) or under duress (*beones*), she is permitted to him, as it is said, "neither she be taken" [Num. 5:13], implying that if she were seized, she is permitted... the same rule applies both to the woman who acts inadvertently and to the one who acts under duress, because an inadvertent act has an element of duress in it.[21, 22]

4. Inadvertent Adultery

What role does inadvertent error play in adultery? What does Jewish law view as unwitting adultery, in which case the adulterous woman is permitted to remain with her husband, and what does it perceive as voluntary action which renders the woman forbidden to both her husband and her lover? Is the role of inadvertent error the same here, in matters of adultery, as it is in other areas of Jewish law? What is the "element of duress" to which Maimonides refers?

These questions will all be addressed. I begin with a discussion of the relevant passages in the Mishnah, Babylonian Talmud and Jerusalem Talmud, then consider the views of the Early and Later Authorities, and a contemporary rabbinical decision on the issue, after which I offer my own analysis.

i. The Mishnah

a. Mishnah Jebamot brings the following case:

> Two men who betrothed two women, and when they were under the marriage canopy they exchanged each one's bride for the other's – then both are guilty under the law of the married woman... . And they must be separated for three months lest they be pregnant; and if they were too young to conceive, they are returned immediately.[23]

---

19 See Rashi *ad loc.* s.v. *vehi.*

20 The wife of a priest is forbidden to her husband even if her adultery was due to rape. See bJebamot 56b; Maimonides, *Code*, Laws Concerning Marriage 24:21; Laws Concerning Forbidden Intercourse 18:7–8; Tur, Even Haezer 6:32–35; *Shulhan Arukh*, Even Haezer 6:11.

21 Maimonides, *Code*, Laws Concerning Marriage 24:19–20.

22 This opinion is accepted by the Tur at Even Haezer 6:34 and by the *Shulhan Arukh*, Even Haezer 6:11.

23 Mishnah Jebamot 3:10, bJebamot 33b.

Two couples had a double marriage ceremony and the women were exchanged[24] for each other.[25] The Mishnah rules that those involved must bring a sin offering (*hatat*) for their inadvertent transgression of the prohibition against adultery. In addition, if the women are past puberty, each couple must be separated for three months to determine if the wife has become pregnant from the intercourse that took place, since any children of this union would be *mamzerim*. If the separation did not take place, it would be impossible to tell who the real father was, and any children would be presumed to have been fathered by the husband.[26]

The conclusion that follows from this Mishnah is that a woman who has sexual intercourse under the erroneous impression that her partner is her husband, is permitted to remain with her husband, and is not deemed a married woman who fornicated with another man.[27] (The temporary separation is, as we noted, only mandated to determine the paternity of any offspring who may have been conceived.)

ii. The Babylonian Talmud

bNedarim relates the following incident:

> A certain woman would rise when sexual relations had taken place and wash her husband's hands. One day she brought him water with which to wash. He said to her, But nothing has happened today. If so, she replied, one of the non-Jewish perfume sellers who were here today – if not you, perhaps one of them.[28]

It was the practice of a certain couple that if they had sexual relations at night, the next morning the woman would bring her husband water to wash

---

24 bJebamot 33b explains the term "exchanged" used in the Mishnah as "were exchanged," i.e., inadvertently and not deliberately.

25 Although such incidents are rare, they do occur, especially in marriage ceremonies where the husband and wife are not acquainted prior to the wedding. Compare Gen. 29:23–25, where Leah was exchanged for Rachel on Jacob's wedding night.

26 See Rashi *ad loc.* s.v. *umafrishin otan.*

27 Compare the position taken in *Yam Shel Shlomo* regarding Jebamot 3:17 – "in any event they are permitted to each other after three months, they are as women who were raped, for they were inadvertently exchanged." Here too, we see that error is associated with duress. (The continuation of the passage will be analyzed below at length.)

28 bNedarim 91a–b.

his hands. One morning, when she brought her husband water to wash his hands, he was taken aback, saying that since they had not had relations the previous night, he was puzzled as to why she had brought him water to wash. She replied that if that was the case, the man with whom she had had intercourse during the night must have been one of the non-Jewish perfume vendors,[29] since if it was not her husband, it could only have been one of them.[30] The Gemara continues: "R. Nahman says: she had a passion for another."[31]

According to R. Nahman, the woman is not to be believed, for she may have coveted another man and made up the story in order to gain a divorce from her husband. R. Nissim Gerondi[32] comments:

> A certain woman – she was the wife of a priest. For had she been the wife of an Israelite, then since, according to her story, she thought it was her husband, and there is no clearer case of duress than this, she would have been permitted to her husband, and there would be no reason for R. Nahman to say "she had a passion for another." But it is as we said.[33, 34]

The Early Authorities understood the incident discussed in the Gemara to involve the wife of a priest, for otherwise R. Nahman's statement would have been superfluous. Were the woman the wife of an Israelite, it would not matter whether or not her story was true, for in any case it would not be grounds for divorce.[35] R. Nahman's statement makes sense only if we interpret the

---

29 See Rashi *ad loc.* s.v. *ahloai* (dealers in aloe).

30 This type of incident, too, though rare, does occur. A similar case of mistaken identity took place in England: A woman was in bed when a man whom she thought was her lover passed by the window. She invited him in, had relations with him in the dark, and only afterward discovered that he was a stranger. The man was charged with the criminal offense of burglary, but acquitted at trial. See Reg. v. Collins [1972] 3 *WLR* 243; cf. 88 *LQR* (1972), pp. 458–459; 28 *Hapraklit* (1973), p. 452.

31 bNedarim, *loc. cit.*

32 R. Nissim b. R. Reuben Gerondi (Ran) was a prominent halakhic authority in fourteenth-century Spain. He wrote novellae and commentaries on the Talmud as well as responsa.

33 *Loc. cit.* s.v. *hahi ita.*

34 R. Nissim Gerondi's explanation accords with the generally accepted understanding of the passage. See, for example, *Tosafot ad loc.* s.v. *shema eineha natna beaher* (she may have coveted another man); Rosh *ad loc.* (in the Vilna edition, the commentary is adjacent to the text) s.v. *ahloai.*

35 See note 34.

incident as having to do with the wife of a priest, for the wife of a priest who commits adultery, even under duress, is forbidden to her husband, and he must divorce her.[36] Thus there is reason to suspect the woman of fabricating the story in order to force her husband to divorce her and enable her to marry someone else.

It is clear from this discussion that a woman who has relations with someone whom she mistakenly thinks is her husband is permitted to remain with her husband, if he is not a priest, for "there is no clearer case of duress than this."

The incidents described in Mishnah Jebamot and bNedarim were not explicitly labelled there as cases of inadvertent transgression. The passage from the Jerusalem Talmud to which we now turn, however, does indeed specifically cite the law that a woman who commits adultery inadvertently is permitted to remain with her husband. But it does not explore whether this refers only to a woman who errs physically – who has intercourse because she is under the impression that her partner is her husband, or also to the woman who is unaware that it is forbidden to fornicate with another, i.e., to one whose mistake is a mistake of law.

iii. The Jerusalem Talmud

The following passage is found in the Jerusalem Talmud:

> He acted deliberately and she inadvertently – it is clear that she is permitted to her house.[37] If her husband divorces her, is [the lover] permitted to marry her? Is it possible that he acted deliberately, and you say it is permitted? If he acts unwittingly, and she deliberately – it is clear that she is forbidden to her house. If her husband divorces her, is [the lover] permitted to marry her? Is it possible to say that she leaves because of him and you say it is permitted?[38]

The Jerusalem Talmud deals with two cases of adultery. In the first case, the male partner commits adultery deliberately and the female inadvertently. In this case the woman is permitted to remain with her husband, but if her husband divorces her, she cannot marry her lover, for he acted deliberately.

---

36 See above, note 20.

37 I.e., to her husband. The *Korban Haeida ad loc.* s.v. *hu mezid vehi shogeget*, comments: "it is clear that she is permitted to her husband if he is not a priest, for just as a wife who was raped is permitted to him, so is the wife who erred inadvertently."

38 jSota 5:1 (20a).

In the second case the positions are reversed: the male partner acts inadvertently and the woman commits adultery deliberately. Here, the law is that the woman is forbidden both to her husband and to her lover, though he erred inadvertently, since it is because of the act she committed with him that she must be divorced.What is the source for this ruling? The passage continues:

> And how do we know that the matter depends on her? Simon bar Ba says in the name of R. Johanan: it is written: "And thou shalt not lie carnally with thy neighbor's wife, to defile thyself with her";[39] on her[40] the matter depends: if she acted deliberately, she is forbidden, if she acted inadvertently, she is permitted.

In summary, we find in the Talmud, in bJebamot and bNedarim (the case in the Jerusalem Talmud is not detailed enough to be relevant here), two cases in which a married woman who inadvertently committed adultery is permitted to remain with her husband. In both cases, the error is factual; the woman erred regarding the identity of the man involved, innocently believing that the man with whom she had sexual relations was her husband, although it subsequently became clear that he was a stranger.[41] In these cases the law is that

---

39 Lev. 18:20.

40 R. Baruch Halevi Epstein comments that the exegesis turns on the phrase "with her" in the biblical verse, which is ostensibly superfluous, and therefore interpreted as "all depends on her." See *Tora Temima*, Lev. 18:20, n. 54; cf. *Pnei Moshe* on Lev. 21:2 s.v. *letama ba*.

41 An additional example of error in fact is mentioned by Rashi, commenting on Sota 28a s.v. *sota lo asa ba*:

> And it is written: "neither she be taken" – she is forbidden, she was taken – she is permitted [see above, near note 16]. This is the law for the inadvertent transgressor. His wife and another married woman are in his home, he meant to be with his wife, and happened on the other woman, and she too erred and thought he was her husband.

In Rashi's example both the man and the woman act in error, as in the case in Mishnah Jebamot. However, even when only the woman acts inadvertently, and the male partner acts deliberately, the woman is permitted, as we saw in the cases in bNedarim and the Jerusalem Talmud. This being the case, why did Rashi use this example, from which it could be inferred that the man involved must also act inadvertently? The *Brit Yaakov*, Even Haezer 94 s.v. *hen emet*, raises but does not answer this question. His views on inadvertent adultery in general will be considered below.

the woman is permitted to remain with her husband, just as she is in the case of rape. In other words, a mistake of fact leading to an act of adultery is considered an inadvertent error, so the woman is permitted to remain with her husband.

The question is, does mistake of law in the context of adultery also constitute inadvertent error, as it does with regard to laws concerning the Sabbath and the law against eating prohibited fat? There is no explicit discussion of the boundaries of inadvertent adultery in the talmudic literature; hence, we must turn to the post-talmudic literature in an attempt to resolve this question.

iv. The Early Authorities

We now examine two responsa by Early Authorities which constitute the basis for all further discussion of the issue in the later halakhic literature.

a. The Responsum of R. Solomon b. Adret[42]

The following problem was presented to Adret:

> A woman was affianced to Reuben, and this Reuben dined at the house of his future father-in-law and requested that his affianced bride be seated with them at the table. Her father did not want to let her sit among them without Reuben giving her a ring, an act considered by them to be as marriage (*kidushin*); for this is their custom. Reuben and the father of the bride reached an agreement, and she was not a child. She was brought before them and Reuben gave her a ring in front of all those assembled. ... after a while something happened between them and she went and married Simon, without Reuben's having given her a divorce, and she had children with Simon. Then Reuben came and brought witnesses to the fact that he had given her a ring and married her, and the witnesses testified that he had given her a ring in their presence.[43]

---

42 R. Solomon b. Adret (Rashba), 1235–1310, a leader of Spanish Jewry in the thirteenth century, was one of the greatest writers of responsa in halakhic history. He also wrote commentaries and novellae on the Talmud, as well as legal codes, the structure and systematic categorization of which served as models for later halakhic literature. (On Adret's system of codification, see n. 100 below and the sources cited there.)

43 *Responsa*, part 1, #1189.

Following this description[44] of the events, Adret was asked, among other things,[45] the following question:

> Also inform us: if the first one now divorces her, is she permitted to the second? Is she considered as having committed adultery under duress and therefore permitted to the first? And since she is permitted to her husband, is she then permitted to her partner; or do we say that she is forbidden to both?

In other words, in this case, where the woman commits adultery under the impression that she is not married to her first husband, is she considered to have acted inadvertently, or is she considered a deliberate adulteress who is forbidden to both her husband and her lover?

Adret replies:

> If one says, this is a case of duress for she did not know that she is forbidden to marry – this is not so, for she should have checked, and since she did not check she is forbidden to both. ... but what sort of duress was this that she can rely on it to marry? If so, we will permit all women who commit adultery, saying: she did not think she was forbidden to do this. And this is clear.

Halakhic scholars are divided on the explanation of Adret's reasoning, and we will discuss their views below in reviewing later halakhic opinion on inadvertent adultery. But first I wish to consider R. Joseph Colon's responsum, which, like that of the Rashba, served as the basis for subsequent halakhic analysis of the problem.

b. The Responsum of R. Joseph Colon

R. Joseph Colon[46] was asked the following question by R. Jacob Landau:

> And the question asked by the Maharil (may the Lord protect him and give him life) concerning a woman who committed adultery voluntarily, but did not know whether or not it was prohibited, is this considered inadvertent transgression?[47]

---

44 This case is quoted in detail because of the relevance of these details to the discussion which follows. However, details concerning the relatives who were present at the ceremony have been omitted.

45 Another question involved relatives who were present at the ceremony.

46 R. Joseph b. Solomon Colon (Maharik), a preeminent scholar in Italy in the fifteenth century.

47 "Maharil," is most likely R. Jacob Landau, author of the *Agur*, and a contemporary of Colon. In other responsa he is referred to as Mahari Landau or Mahari

Colon responds:

> In my humble opinion this is not considered a case of inadvertent transgression, where she is permitted to remain with her husband, for she intended to break faith with her husband and commit adultery. For it is not written "a man whose wife goes astray, and breaks faith with the Almighty," meaning that she transgresses intentionally, but it is written, "and she breaks faith with him."[48] What Maimonides wrote in Laws Concerning Marriage, ch. 24,[49] that a married woman who inadvertently, or under duress, fornicated with another, is permitted to remain with her husband, refers to unwitting error as regards the physical act of adultery, but not error as regards the prohibition, for that is not considered unwitting adultery, for she intends to commit adultery, but does not know that it is prohibited. And what is considered unwitting adultery? For example, when she believes he is her husband and he turns out to be someone else, as in the incidents at the end of Nedarim.[50, 51]

The interpretation of this responsum and its relation to Adret's responsum were disputed by later halakhic authorities. We now turn to a survey of the different ways in which the responsa were understood.

v. The Later Authorities on the Responsa of R. Solomon b. Adret and R. Joseph Colon

a. R. Joseph Caro

R. Joseph Caro[52] presents Colon' s decision as law in his *Beit Yosef.* I quote the passage in its entirety; note which section of the responsum Caro chooses

---

Lando; see *New Responsa and Rulings of Joseph Colon* (Hebrew), ed. A.D. Pines (Jerusalem: 1970), index of names. On the relationship between Colon and Landau, see *ibid.*, p. 44, n. 151.

48 Num. 5:12.

49 This refers to Maimonides's conclusion, quoted above at note 21.

50 This refers to the Gemara at the end of bNedarim quoted above at note 28.

51 In the first edition of Colon's Responsa (Venice, 1518), responsum #168. However, this edition erroneously numbered the responsum following #164 as #166, not #165. In most later editions – Cremona, 1558, Lemberg, 1798, and Sedilkov, 1834 – the responsum is #167. In the 1884 Warsaw edition, [photo offset (New York: 1958 and 1968)], it is erroneously labelled #165.

52 R. Joseph b. Ephraim Caro, 1488–1575, one of the greatest halakhic authorities of all time. Author of the *Beit Yosef, Shulhan Arukh* (see below at n. 100), *Kesef Mishne*, a commentary on Maimonides, and a book of responsa entitled *Avkat Rokhel.*

to quote:

> And he wrote[53] in #168:[54] a woman who committed adultery voluntarily without knowing that it was prohibited, is not considered to have erred inadvertently so that she is permitted to remain with her husband. And as to what Maimonides wrote in ch. 24 of *Laws Concerning Marriage*[55] – that a married woman who inadvertently, or under duress, fornicated with another, is permitted to remain with her husband, that is, if, for example, she thinks it is her husband but he turns out to be someone else; he brought arguments proving it.[56]

Before discussing the sentence which follows this passage, I want to point out that Caro does not include Colon's homiletic exegesis of the verse "and she breaks faith with him" in his extract, mentioning only Colon's decision and his interpretation of Maimonides.

Does Caro take Adret's responsum to be in agreement with Colon's opinion that the category of inadvertent adultery covers only the case of one who is mistaken about the identity of her husband, and not that of one who voluntarily commits adultery in ignorance of the prohibition? If indeed the latter case is not covered, such women are not permitted to remain with their husbands.

This certainly seems to be the case, for at the end of the passage quoted above, Caro states "and this is the import of Adret's responsum #1189." However, careful scrutiny of the editions of the *Beit Yosef*[57] reveals that this

---

53 The reference is to Colon, *loc. cit.*; see the paragraph which precedes the passage quoted.

54 Caro used the Venice edition of Colon's responsa, in which the responsum is #168 (see note 51).

55 See the text above near note 21.

56 *Beit Yosef* on the Tur, Even Haezer, sec. 115, s.v. *vema shekatav rabeinu*, and where she tells him she committed adultery, in the middle of the passage.

57 In the first editions of the *Beit Yosef* the passage ends with the words "And he brought arguments proving it," with no reference to Adret's responsum; see the Adelkind edition, Sabbioneta, 1553. The same is true for the two Venice editions of 1565, Borgo di Cavalli (where the year of publication is given as – "*shnat rav lanu od* – Caro lives": in *gematria*, the words "Caro is yet alive" comes out to 1565), and Juan di Gara, printed in Bombirghi type, as well as the Cracow edition of 1613. The first edition in which the sentence referring to Adret's responsum appears is the Berlin edition of 1702, the first to print both the *Beit Yosef* and the *Darkei Moshe* commentaries on the Tur (see note 59). In the Wilhermsdorf edition of 1727 (which brings, in addition to the *Beit Yosef*

passage is not authentic. It apparently originated in Isserles's *Darkei Moshe*[58] and was interpolated into Caro's commentary.[59]

In any event, Caro accepted Colon's decision and cited it in the *Beit Yosef*. However, the *Shulhan Arukh*, also written by Caro, does not mention the ruling, and as we will see below, this omission occasioned the claim that Colon's decision is not to be followed in practice.[60]

---

and the *Darkei Moshe*, yet another commentary, *Haprisha Vehadrisha*), this sentence appears in the *Beit Yosef*. A different state of affairs emerges in the Slavuta edition of 1771: the sentence referring to Adret's responsum, "and this is the import of Adret's responsum #1189," appears, not in the *Beit Yosef*, but in the *Darkei Moshe*, par. 15, though it had not appeared in previous editions of the *Darkei Moshe*. It is interpolated between the original paragraphs 14 and 15, with the original 15 becoming 16, and the other paragraphs renumbered accordingly.

Of editions printed after 1795, some cite the sentence about Adret in the *Beit Yosef* and others in the *Darkei Moshe*: the Dyhernfurth edition of 1795 has the sentence in the *Darkei Moshe*, while in the 1801 Slavuta, 1813 Vienna, and 1836 Capusta editions, the reference to Adret is found in the *Beit Yosef*, and there is no mention of it in the *Darkei Moshe*. In the 1859 Zhitomir, 1861 Lvov, 1864 Warsaw, and 1900 Vilna editions, the sentence is cited in the *Darkei Moshe* only and does not appear in the *Beit Yosef*.

58 *Darkei Moshe* is the commentary on the Tur written by R. Moses Isserles (Rema), 1525–1572. As well as responsa, Isserles, a preeminent halakhic authority, also wrote the *Mapa* commentary on the *Shulhan Arukh* [*mapa* means tablecloth; *shulhan arukh*, prepared table].

59 Our argument that this passage was not written by Caro is supported by the fact that it is not found in the earlier, and presumably more authentic, editions of the *Beit Yosef*. Isserles is the likely author of the claim that Adret's responsum indicates that his ruling is the same as Colon's. It seems that the first editions mistakenly interpolated the sentence into the *Beit Yosef* and only later restored it to its correct location, the *Darkei Moshe*. The fact that it appears for the first time (in the *Beit Yosef*) in the 1702 Berlin edition, the first edition in which the *Darkei Moshe* appears, would seem to support this hypothesis. Before the printing of *Darkei Moshe* I can find no version where the passage in question concludes with the sentence "and this is the import of Adret's responsum #1189." If indeed this premise is correct, the variations in the later editions are the result of the publishers' reliance on the first editions: publishers who used the 1702 Berlin edition, or the 1727 Wilhermsdorf edition, cited the sentence as part of the *Beit Yosef*, while those who worked with the 1771 Slavuta edition cited it correctly. For more on this, see below at notes 101–108.

60 On this claim and the response to it, see below, near notes 93–100.

b. R. Moses Isserles

As we noted, R. Moses Isserles[61] states that it is not just Colon's opinion that a woman is forbidden to remain with her husband if, unaware that it is prohibited, she commits adultery, but that this opinion is also expressed in Adret's responsum. A number of other halakhic authorities, however, felt that the responsa in question are not in agreement on the matter.

The *Shulhan Arukh*, as we pointed out,[62] mentions neither the responsum of Colon nor that of Adret. In his remarks on the *Shulhan Arukh*, Isserles brings the ruling in Colon's name alone:

> An adult female who committed adultery inadvertently, thinking the man was her husband, though he was another, is permitted to remain with her husband if he is not a priest (Maimonides, *Code*, Laws Concerning Marriage, ch. 24).[63] However, an adulteress who believed it permissible to commit adultery is considered to have acted deliberately and is forbidden to her husband even if he is not a priest[64] (Colon, #168).[65, 66]

Isserles does not mention that this ruling can be attributed to Adret as well. Apparently, he considered his statement in *Darkei Moshe* sufficient.

c. R. Solomon Luria

R. Solomon Luria of Poland,[67] a contemporary of Caro and Isserles, also writes on the case of the woman who commits adultery inadvertently. Luria

---

61 See note 58 above.

62 See above at note 60.

63 This passage from Maimonides is quoted above at note 21. Although Maimonides used the general term, *shogeg*, Isserles, following Colon, understood Maimonides to refer only to error of fact. Undoubtedly, Maimonides' statement that "the same rule applies both to the woman who acts inadvertently and to the one who acts under duress because an inadvertent act has an element of duress in it" (Laws Concerning Marriage 24:20) contributed significantly to this interpretation.

64 Given that a priest's wife is forbidden to remain with him even if she mistook another man for her husband, it goes without saying that she is forbidden if she was unaware that adultery was prohibited, since with regard to priests, it is the physical act of intercourse that makes a wife forbidden to her husband; see note 20 above.

65 Isserles also had the Venice edition of Colon's responsa. See note 51 above.

66 Isserles's commentary on the *Shulhan Arukh,* Even Haezer 178:3.

67 R. Solomon b. R. Jehiel Luria (Maharshal), 1510–1570, author of *Yam Shel Shlomo* and a commentary on the Talmud.

first cites *Or Zarua*,[68] then states his own opinion, in light of which he offers a strict interpretation of the opinions of Maimonides and the *Sefer Mitzvot Gadol*. At the end of his remarks, the responsa of Adret and Colon are quoted. Luria's view is as follows:

> The *Or Zarua*[69] states: the wife of one who is not a priest who cohabited voluntarily with another man and thought him to be her husband, is considered to be a case of adultery under duress[70] and she is permitted to her husband … .[71]
>
> And he ruled correctly.
>
> In any event, only something of this sort would be considered like rape. But if she acted voluntarily, and thought that there was no prohibition, she is forbidden. Maimonides' statement, that she committed adultery inadvertently or under duress, does not mean that she erred about the prohibition, but that she thought him her husband: but if she erred and thought there was no prohibition, it is clear that she is forbidden, for in the final analysis she is an adulteress and forbidden.
>
> Accordingly, we must aver that Maimonides' statement in the *Code*, Laws Concerning Marriage,[72] and similarly, the assertion of *Sefer Mitzvot Gadol*,[73] that

---

68 *Or Zarua*, by R. Isaac b. R. Moses of Vienna, a thirteenth-century halakhic authority.

69 *Or Zarua*, Laws Concerning Levirate Marriage and Halisah, 726. The *Or Zarua* refers to the passages in the Mishnah and the Gemara discussed above, which deal with the case of the women who were exchanged (see above at note 23ff.). The reason I cite the *Or Zarua* here, in the context of Luria's statement, and not, as would be chronologically correct, at the beginning of the section on post-talmudic literature, is that, probably as a result of the *Or Zarua*'s limited circulation, it seems to have been unknown to later halakhic figures who studied the matter. The remarks of these authorities thus focused only on analysis of the responsa of Adret and Colon.

70 Here too we see the link between rape and inadvertent error. R. Isaac of Vienna is making the point that the woman is permitted to remain with her husband and the act does not incur the consequences of adultery; he compares the woman who acts inadvertently to the woman who is raped, apparently on the basis of the interpretation of the term "taken" (see above at note 18).

71 The *Or Zarua* goes on to discuss the problem of the status of a child born of this act, a question which does not concern us here. Compare the discussion above in note 5, sec. (2).

72 See above at note 21.

73 *Sefer Mitzvot Gadol* (Semag), positive commandments, 48 (near the end; in the Venice edition, 1547, p. 128a). The *Sefer Mitzvot Gadol*, a codification of the law in which subjects are arranged by ordering the positive and negative commandments, is the work of R. Moses of Coucy, a thirteenth-century scholar.

> the married woman who fornicated with another, either inadvertently or under duress, is permitted, is based on the "Young Woman" chapter in the Gemara, which states "neither she be taken,"[74] comparing the woman who commits adultery inadvertently to one who is raped.[75] We say this is where she erred and believed she was not married; even if she cohabited voluntarily, it is considered to be under duress. And so Adret wrote in a responsum concerning Michal, the daughter of Saul. But Adret did not write about a woman who committed adultery not knowing that there was a prohibition against it. However, I found an opinion much like my own in Colon's responsum #160:[76] this is not considered a case of inadvertent transgression, where she is permitted to remain with her husband, for she intended to break faith with her husband and commit adultery. For it is not written "a man whose wife goes astray, and breaks faith with God," meaning that she transgresses intentionally, but it is written, "and she breaks faith with him." He went on to bring much evidence for this view, but this is the point I wish to make.[77]

After quoting the *Or Zarua*, Luria explains his own view, namely, that the woman is permitted to remain with her husband only in cases of mistake of fact. The woman who does not know that it is prohibited to commit adultery is forbidden to her husband, because in the final analysis, she committed adultery. When he goes on to explain the opinions of Maimonides and the *Semag*, Luria uses as an example the case of the woman who mistakenly believed that she was not married. It is hard to understand why he does not use the example to which he referred previously, that of the woman who had sexual relations with one she mistakenly thought to be her husband. Had Luria used it, this example would have made it clear that there is a definite difference between mistake of fact and error due to ignorance of the law. Luria's example is problematic, for it is difficult to pinpoint the difference between the case of a woman who is ignorant of the law against adultery, and that of one who is ignorant of the laws of marriage, believing that her

---

74 Luria refers to the discussion in the Gemara quoted above at note 18.

75 This is the correct wording. The printed editions cut off the feminine gender ending (*a*) of the word for "woman who has been raped" (*anusa*), leaving the masculine "one who has been raped" (*anus*), but it is clear, given that the sentence is about a woman, that this should be emended to "*anusa.*"

76 None of the extant printed editions list the responsum as #160. A letter was most likely erased [in the Hebrew numbering system, letters represent digits], and the number should read either 168 or 167 (see note 51 above).

77 *Yam Shel Shlomo,* Jebamot 3:17.

husband is not legally married to her, when in fact, he is. This point will be taken up at the end of the paper.

To support his example, Luria cites Adret's responsum concerning Michal the daughter of King Saul. Adret wrote two responsa that refer to this episode. In the first, which is devoted to a discussion of the case of Michal – her marriage to David, her marriage to Palti, and her return to David – Adret states:

> Therefore, if Michal believed that she was permitted to marry Palti, because her marriage to David was invalid, as Saul believed, this is considered to be a case of duress, and hence, she is permitted to return to him [David].[78]

Luria apparently referred to this responsum. However, Adret addressed the matter of Michal a second time, while considering the case, cited above, of the woman who believed her first marriage invalid and therefore married someone else, circumstances very similar to those of Michal. There, Adret argues as follows:

> If you say: then how was Michal, the daughter of Saul, permitted to David? The two cases are different, for in the case of Michal, she was clearly under duress, for Saul compelled her to marry Palti, and she was forced to have relations against her will, as was Esther with Ahasuerus. Or you can say that it was a mistaken judicial ruling, for Saul's court ruled that "when a man marries a woman with a penny (*pruta*) and a loan, he considers the loan to be the significant factor," and she relied on this ruling, and this sort of error is like duress.[79]

We can infer from Adret's argument that when the second marriage is not against the woman's will or when the error regarding the validity of the first marriage is not an erroneous ruling by a distinguished court, the woman would be forbidden, as, indeed, Adret ruled in this responsum.

We have no choice but to conclude that Luria referred only to the first of Adret's responsa, overlooking the second, which restricts the precedent of Michal the daughter of Saul, citing its unique circumstances. If we adopt Adret's reasoning, an example of inadvertent adultery in which the woman is permitted to remain with her husband would be a case in which she had relations with another man thinking him to be her husband. If, however, the

---

78 *Responsa*, part 1, #10.

79 *Responsa*, part 1, #1189; see above at note 43ff.

woman did not know that she was married, mistakenly believing that her marriage ceremony was invalid, or if she believed that there was no prohibition against adultery, she is forbidden.

In concluding this analysis of Luria's opinion, let me point out that Luria considered the interpretation of the verse "and she breaks faith with him" to be Colon's crucial argument. Below, we will see that Luria smoothed the way for later authorities who perceived the essence of Colon's argument to be his homiletic exegesis of this verse. Several of them proceeded to use it to differentiate between the laws of inadvertent adultery and other laws governing inadvertent transgression.

d. R. Judah Ayash and R. Haim Joseph David Azoulay

Colon's novel understanding of the law, which distinguishes the case of the woman who inadvertently commits adultery from other cases of inadvertent transgression in Jewish law, was the subject of a dispute between two halakhic scholars in the eighteenth century. R. Judah Ayash[80] attacked Colon's opinion, while R. Haim Joseph David Azoulay[81] came to its defense. For clarity and convenience, I consider both opinions together.

Ayash's first objection[82] is that Colon derives his opinion from his own homiletic exegesis of the biblical verse "and she breaks faith with him," yet

80 R. Judah b. R. Isaac Ayash, c. 1700–1760. An eminent Algerian scholar, Ayash died in the land of Israel. *Lehem Yehuda*, his commentary on Maimonides' *Code*, was published in Leghorn in 1745. His discussion of Colon's responsum appears in *Lehem Yehuda*, Laws Concerning Marriage, ch. 24.

81 R. Haim b. R. Joseph David Azoulay (Hida), 1724–1806. A halakhic authority who spent much of his life in Israel, he wrote the *Haim Shaal* responsa, *Shem Hagedolim, Birkei Yosef*, and numerous other works. For more on Azoulay, see M. Benayahu's comprehensive *R. Haim Joseph David Azoulay* (Hebrew), (Jerusalem: 1959).

82 Ayash prefaces his critical review of Colon's opinion with remarks acknowledging his audacity in disputing the views of such an eminent predecessor:

> Since the crown of the Torah lies in a corner, and whoever wishes to take it up, may do so, I have come to examine the details of this responsum as a student who speaks before his teachers prostrated upon the ground, for in my view, that of the layman, the opinion of the Rabbi was beyond my understanding, in terms of both the reasoning and the evidence (*Lehem Yehuda*, p. 42b, col. 1).

this procedure was permitted only until the end of the talmudic period.[83] Ayash argues as follows:

> I was astounded that a holy mouth could say this, for after the redaction of the Talmud, no scholar can derive any law whatever from his own exegesis, especially when it is done to make something permissible. Since the Talmud did not offer this exegesis, or take special note of the grammatical point on which it is based – we cannot add or subtract.[84]

Azoulay,[85] citing another of Colon's responsa,[86] in which Colon states unequivocally that since the redaction of the Talmud the homiletic exegesis of biblical verses has been expressly forbidden,[87] explains that Colon did not intend to engage in exegesis:

> This type of exposition is not considered exegesis but rather explication of the verse, for even though she erred in the law, thinking it was permitted to commit adultery, she certainly knew that she was betraying her husband by desiring another man. Therefore, although she believed it permissible to commit adultery, any woman who breaks faith with her husband and fornicates with another is forbidden to him. For the verse states, "and she breaks faith with him [Num. 5:12]," hence, regarding a woman who betrays her husband, her betrayal has legal consequences.[88]

Ayash presents an interesting argument which penetrates to the heart of the problematics arising from the laws of mistake of fact and mistake of law, and

83 For a detailed discussion of the homiletic exegesis of biblical verses, see M. Elon, *Jewish Law* (Philadelphia and Jerusalem: 1994), vol. 1, pp. 275ff. On the problematics of Midrash in the post-talmudic period, see *ibid.*, pp. 391ff. In discussing this issue, Elon cites the arguments developed in the Hebrew version of this paper, see *ibid.*, p. 399.

84 *Lehem Yehuda ad loc.*

85 In his *Haim Shaal* (Leghorn, 1789; Lemberg, 1886; Brooklyn n.d. [photo offset by B.Z. Hacohen]), part 2, #48.

86 Responsum #139 in the Venice, 1518; Cremona, 1558; Lemberg, 1798; and Sedilkov, 1834 editions. In the 1884 Warsaw edition it appears as #138.

87 Colon's words are as follows:

> And moreover, you sit and expound the following exegesis: "and her *yabam* shall come to her" – and not to another – and so on. This exegesis is mentioned neither in the Talmud nor in the halakhic authorities, and you know that we are not allowed to come up with our own exegeses.

88 *Haim Shaal,* p. 2b, col. 1.

notes the injustice suffered by one who makes a mistake of law, or is ignorant of the law:

> Moreover, how can we imagine that this woman who acts inadvertently and does not know the prohibition, will be sentenced to "and her belly shall swell"[89] and so on? Does the Almighty judge the person who acts inadvertently as he does one who acts deliberately?[90]

Azoulay responds as follows:

> This is not difficult, for even if she claimed that it is permitted to commit adultery, she deliberately broke faith with her husband, and for that she is punished.[91, 92]

Here I will hint at the thrust of the argument I develop at the end of the article. As Ayash rightly points out, the Almighty does not judge the unwitting transgressor as the deliberate transgressor is judged, and from the divine perspective, there would be no justification for doing so, since the unwitting transgressor has no intention of wronging the Almighty. Therefore, the woman who committed adultery in ignorance of the law may, from the divine perspective, be considered an inadvertent transgressor, just as one who inadvertently transgresses by doing forbidden work on the Sabbath does not know that his action is forbidden. However, in interpersonal relationships, to consider transgression arising from ignorance of the law to be inadvertent is out of the question, since if this were the case, every individual would set his own standards of conduct as he saw fit, and his fellow would be at a loss. In the case before us, this would be the husband whose wife "betrays" him and who would not be allowed to divorce her. Therefore the law states that inadvertent transgression against the husband due to ignorance of the law constitutes betrayal, and the husband is permitted to divorce his wife.

---

89 According to biblical law, the wayward woman (*sota*) had to drink "the water of bitterness that causeth the curse." If she had committed adultery, her belly would swell and her limbs would "fall away"; see Num. 5:11, 21–22 and 27. And see H.H. Cohn, "Ordeal," in *Principles*, p. 524.

90 *Lehem Yehuda, loc. cit.*

91 On the question of whether in such a case she drinks the bitter waters and whether the ordeal is intended to test her, see *Mishne Lamelekh*, Laws Concerning the Wayward Woman, 2:8 s.v. *mikol makom* (2nd occurrence) and the rest of the responsum in *Haim Shaal.*

92 *Haim Shaal, loc. cit.*

I will not discuss all the arguments and responses in the *Lehem Yehuda* and *Haim Shaal*, but will conclude my review of these opinions by raising two additional points, the first, because of the broader question it raises, and the second to corroborate my hypothesis about the different versions of Caro's interpretation of Adret's responsum in the texts of the *Beit Yosef* and *Darkei Moshe*.

Ayash ends his arguments against Colon with these words:

> In my humble opinion, it is for this reason that the author of the *Beit Yosef* omitted this responsum and did not copy its ruling into either the *Beit Yosef* or his short book.[93] However, in his commentary,[94] R. Isserles, of blessed memory, was swayed by Colon and wrote: In my opinion one should not rely on this even in practice.[95]

As we saw,[96] Colon's opinion is indeed quoted in the *Beit Yosef*, and it is difficult to understand how Ayash overlooked it.[97] Ayash was so certain of the correctness of his opinion that he even stated that it applied in practice.[98] Azoulay calls Ayash's attention to the omission and explains the fact that the *Shulhan Arukh* does not cite this law as follows:

> And the most important point is that the author of the *Lehem Yehuda* overlooked the fact that the *Beit Yosef*, Even Haezer, sec. 115, quoted this responsum of Colon's, and wrote that Colon brought arguments proving his point. It appears

---

93 This refers, of course, to the *Shulhan Arukh*, a summary of the *Beit Yosef*.

94 Isserles's remarks are quoted above at note 66.

95 *Lehem Yehuda*, p. 43a, col. 1.

96 See above at note 53ff.

97 Ayash also overlooked Adret's responsum, part 1, #1189 (see above), as Azoulay states: "and R. Ayash overlooked Adret's responsum in his *Lehem Yehuda*, when he ruled contrary to the opinion of Colon, even in practice" (*Haim Shaal*, p. 3a, col. 2). On Ayash's decision in practice, see the next note.

98 Here Ayash contravenes the almost universal policy among the Later Authorities of avoiding overt disagreement with the Early Authorities. Even after presenting numerous arguments to counter those of the Early Authorities, for example, they end their discussions with the disclaimer that they do not intend their opinions to be applied in practice, and only if other contemporary authorities concur with their opinion, or precedents can be found in other works by Early Authorities, would they dare decide the law in practice in a manner incompatible with the Early Authorities' rulings. Yet legally, that is, from the halakhic point of view, they are certainly empowered to do so, given the principle: "the law is in accordance with the view of the later authority" (*hilkhata kebatrai*). See Elon, *Jewish Law*, vol. 1, pp. 267ff; and cf. chapter 4 of this volume, by Israel Ta-Shma, "The Law is in Accord with the Later Authority – *Hilkhata Kebatrai*: Historical Observations on a Legal Rule."

> from his matter-of-fact phrasing and from the way in which he added "and he brought arguments proving it" that he agrees with Colon. If someone should say: if so, why did he not include this opinion in the *Shulhan Arukh*?, go and say to him: We have found several instances where incontrovertible laws which he mentioned in the *Beit Yosef* are omitted from the *Shulhan Arukh*, as all those who are well-versed know. Later authorities generally say that this is because it was a simple law in his eyes, or one rarely applied. And we also aver that both considerations are the case here: it seemed simple enough to him, that a woman who believed it was permitted to commit adultery does not constitute a case of inadvertent transgression, and thus, it seemed clear that she is forbidden to her husband. And secondly, a woman's believing adultery to be permissible is not a common occurrence, for all religions, even those of the nations, prohibit it, and it is most improbable, and for this reason the *Shulhan Arukh* omitted it.[99]

Elon's thorough study of methods of codification in Jewish law analyzed Caro's system of codification in the *Beit Yosef* and the *Shulhan Arukh*.[100] His research sheds considerable light on Azoulay's remarks. It seems, as he asserts, that it cannot be assumed that a decision rendered in the *Beit Yosef* and omitted in the *Shulhan Arukh* is invalid. The very fact of its inclusion in the *Beit Yosef* is evidence that Caro considered the law in question valid and obligatory. Its absence from the *Shulhan Arukh* is adequately accounted for by the reasons Azoulay gives in *Haim Shaal*.

The second point I wish to raise in connection with Azoulay's remarks has to do with the discrepancy between the texts of the *Beit Yosef* and *Darkei Moshe* as to the thrust of Adret's responsum. Azoulay states:

> In the Berlin edition[101] of the *Beit Yosef*, which is before me, sec. 115, when the responsum of Colon is cited, the statement "and this is the import of Adret's responsum #1189" appears between half-moons.[102] This seems to be an interpolation, and not the words of our master.[103]

Thus Azoulay was working with the 1702 Berlin edition, where the statement about Adret does indeed appear in parentheses. Azoulay's suspicion that

---

99 Azoulay, *op. cit.*, p. 3a, col. a.

100 See Elon, *Jewish Law*, vol. 3, pp. 1309ff. Caro was not the first to produce a dual codification, that is, one featuring both a detailed version and a general summary, as his work was preceded by Adret's *Torat Habayit Haarokh* and *Torat Habayit Hakatzar*. See *ibid.*, pp. 1312–1313.

101 The reference is to the 1702 Berlin edition, mentioned in note 57 above.

102 Crescents – i.e., parentheses.

103 Azoulay, *op. cit.*, p. 3a, col. 2.

it was an interpolation may perhaps have been based on its content, since he was of the opinion that Adret's responsum was incompatible with that of Colon, as I explain below. Or perhaps earlier editions of the *Beit Yosef* in which the passage did not appear may have made an impression on him. The second explanation seems unlikely, however, given that, were this the case, Azoulay would probably have mentioned it just as he informed his readers that he was using the 1702 Berlin edition. It is clear, however, that Azoulay did not have the version, published only in 1771, in which this sentence appears in Isserles's *Darkei Moshe*.

What was the conceptual incompatibility that led Azoulay to suspect an interpolation? After quoting Adret's responsum, Azoulay says: "For it follows from this statement that a woman who committed adultery and believed that it was not prohibited, is permitted according to law, but we do not say she is permitted, for she should have checked."

It seems to me that Adret's statement concerning the woman's obligation to be careful should be seen as supplementing the real reason – namely, "If so, we will permit all women who commit adultery, saying: she did not think she was forbidden to do this." Hence, the woman is forbidden by law, and not merely as a penalty for carelessness, as I explain below.

e. R. Haim of Volozhin

In his *Hut Meshulash*[104] R. Haim of Volozhin[105] analyzes the question of one who performs an action thinking it is permitted. Unfortunately, a large part of the responsum is lost, and the printed editions are only fragmentary. Nevertheless, it is evident from them that according to Volozhin, Adret's arguments differ from those of Colon, although Adret agrees with Colon[106] that a woman who commits adultery believing it is permissible to do so is

104 Vilna, 1882; photo offset (New York: 1965), par. 13.

105 R. Haim b. R. Isaac Volozhiner, 1749–1821, student of the Gaon of Vilna and founder of the Volozhin yeshiva, where he instituted a widely-acclaimed method of study.

106 Chronologically, Adret precedes Colon, and hence, of course, could not have agreed with Colon's opinion, which was formulated after his death. Thus what is being referred to here is Adret's line of thought. In halakhic discourse the chronological factor is ignored all too often, and one frequently finds discussions about whether a given Amora accepted or disagreed with a position taken by Maimonides, and the like.

forbidden by biblical law to remain with her husband.[107] Volozhin defends this claim as follows:

> And so the *Beit Yosef* writes in sec. 115 after citing Colon: "and this is the import of Adret's responsum #1189."[108]

We have to assume that Volozhin, like Azoulay, was using the 1702 Berlin edition, in which the sentence is presented as part of Caro's *Beit Yosef.*

Although, as mentioned above, textual lacunae make it difficult to reconstruct the details of Volozhin's stand, we will quote its essence in his own words:

> We derive four different rulings from this:
> 1. A woman who commits adultery deliberately, even if she thinks there is no prohibition against it – is forbidden by biblical law, according to both Adret and Colon, each for his own reasons, as I explained above.
> 2. A woman who is married and believed she was not married, even if it was her own mistake, and married another – is forbidden by biblical law according to Adret and by Rabbinic law according to Colon.
> 3. A woman who was married, and was mistakenly told by experts that her marriage was invalid, and married another, and it was then made known that these authorities were mistaken – Adret permits her to remain with her husband and Colon says she is forbidden by Rabbinic law.
> 4. A woman who committed adultery following a mistaken ruling of a distinguished court – Adret permits her to remain with her husband, and Colon considers it forbidden by biblical law.[109]

f. R. Baruch Mordechai Livschitz

R. Baruch Mordechai Livschitz[110] explains the relationship between the responsa of Adret and Colon in a responsum included in his collected responsa *Brit Yaakov*.[111] Like Azoulay, he sees a contradiction between the two responsa. He explains that Adret's statement posits two stages: originally, the woman is permitted to her husband under biblical law, and it was only Rabbinic law that forbade her to him, since she should have been more careful. Livschitz

107 Volozhiner, *op. cit.*, p. 33b, col. 2.
108 *Ibid.*
109 *Ibid.*, p. 33b, col. 2–p. 34a, col. 1.
110 A nineteenth-century halakhic authority in Russia, head of several Jewish communal courts.
111 *Brit Yaakov* (Warsaw: 1897), Even Haezer #95.

understands Colon to say that even under biblical law the woman is forbidden, because of the verse "she breaks faith with him." However, as I suggested above in my comments on the *Haim Shaal*, Adret's statement can be interpreted differently, as I will explain.

vi. A Contemporary Case

Having surveyed the opinions of the Later Authorities on this matter,[112] let us review a contemporary rabbinical court ruling on inadvertent adultery.

The rabbinical court of the Ashkelon district[113] was petitioned by a man and woman who sought permission to marry. The woman had previously been married. On 28 Heshvan, in 1970, a ruling had been handed down granting her a consensual judgment for divorce. Three months later, the woman and her husband were divorced by a writ of divorce (*get*). The court had decisive evidence that the woman had had sexual relations with another man, whom she wished to marry, during the interim period between the court's ruling and her receipt of the writ of divorce. The woman and her partner said they had acted in the belief that it was permissible after the court's judgment had been rendered. That is, they claimed they had transgressed inadvertently, under the mistaken impression that there was no prohibition against their having sexual relations once the court had ordered the granting of a divorce, and that as an inadvertent transgressor the woman was permitted to her husband and her partner; hence, the petitioners should be allowed to marry. Although the court was rather sceptical of the claim that they had begun to have sexual relations only after the court ruling, it decided to take up the challenge and to assess the validity of the petition on the assumption that their story was in fact true.

The court referred to the responsa of Adret and Colon, asserting that the views expressed were indeed contradictory. While for Adret, the court explained, the main point was the woman's responsibility to exercise care, and the prohibition was of Rabbinic origin, since under biblical law the woman is permitted, Colon considered the prohibition to be of biblical origin because of the verse "and she breaks faith with him."[114]

---

112 R. Isaac Kulitz (Kuliditsky) discusses some of the sources analyzed above in his *Minhat Eliyahu* (Jerusalem: 1955), p. 28.

113 The bench was R.I. Goldshmidt (acting head of the rabbinical court), R.M. Lopez and R.M. Militzky. See File 713/1970, Petitioners A and B, RCJ, vol. 8, pp. 184ff.

114 *Ibid.*, p. 187.

In order to conclude that even according to Adret the woman who believes adultery is permissible is prohibited as a matter of biblical law – and thereby explain Volozhin's position as expressed but not justified in the *Hut Meshulash*, apparently due to the fragmentary state of the text – the court proposed a novel way of resolving the issue. The court cites the Gemara at bMakot 7b:

> Our rabbis taught: "by error" – this excludes anyone who kills wilfully; "unknowingly" – this excludes anyone who kills intentionally. "wilfully" – is it not obvious that such a one is deserving of death? Rava said: it is to exclude a case where one says that he thought it was permitted to kill. Abbaye said to him: if he thought that it was permitted to kill, then he is considered a case of duress. Replied Rava: I say that one who says it is permissible is akin to a wilful murderer.[115]

Rava defines the category of inadvertent transgression in cases of murder as excluding cases where the perpetrator says it is permissible, for one who says it is permissible – in our terminology, one who is ignorant of the law – is like one who acts deliberately.

In discussing the laws of murder, Maimonides defines a category between inadvertent murder, where the perpetrator must flee to a city of refuge, and deliberate murder, which is punished by death. This third category is that of inadvertent murder that is virtually deliberate. Such a killer is not exiled to a city of refuge, and must constantly beware of any "avenger of blood" (*goel hadam*), who is permitted to kill him.[116] Among those he includes in this category[117] are "those who believe that it is permissible to slay another."[118] Elsewhere he states: "If a resident alien kills another resident alien, thinking it permissible to kill, he is virtually a deliberate murderer, and must be put to death on account of this killing, since he intended to murder him."[119]

It follows, then, that in the context of the laws of murder, the category of inadvertent transgression does not include ignorance of the law. In this

---

115 The same text is cited at bMakot 9a regarding another matter.

116 Maimonides, *Code*, Laws Concerning Murder and the Preservation of Life 6:4.

117 *Ibid.*, 6:5, where Maimonides brings the most common examples that would fall under this category: murderers who could not be sentenced to death because of some flaw in the evidence brought against them, e.g.: no warning was given, there was only one witness to the murder, etc.

118 *Ibid.*, 6:10.

119 *Ibid.*, 5:4.

regard, the laws of murder differ from the laws of the Sabbath and those of prohibited fat. How is this difference to be accounted for? The *Tosafot*, commenting on tractate Makot, explain it on the basis of exegesis from a verse having to do with the inadvertent murderer: "For here it is different, for *shgaga* is written an extra time in the verse, to exclude one who says it is permitted."[120]

Nachmanides, commenting on the passage, gives another reason:

> One can say that in those contexts it is written "one who transgresses inadvertently," and since he acts in error concerning a transgression that he does not know is forbidden, he is called "one who transgresses inadvertently"; but here it is written: "one who strikes another inadvertently": the actual act of striking is referred to, and since he struck deliberately, he is virtually a deliberate murderer (Nachmanides *ad loc.*).

Hence, asserts the court,

> It follows from this reasoning that the difference is that if it is written that the transgression is inadvertent in cases where one brings a sacrifice, one who thinks it permitted is considered to have transgressed inadvertently, but if it is written, striking someone inadvertently, meaning that he strikes him unwittingly, the unwittingness must be regarding the act of striking itself, and one who thinks it permitted is not considered to have acted inadvertently, for with respect to the act of murder he is considered to have acted deliberately. This is what Colon meant, for the verse reads "and she breaks faith with her husband", not "and she breaks faith with the Almighty." He did not mean to create a new exegesis that is not in the Talmud,[121] but to explain what is considered inadvertent error and what is considered deliberate, making the woman forbidden to her husband. For if the verse read "and she breaks faith with the Almighty" – the meaning would have been, deliberately transgressing, as in the case where one is liable to bring a sacrifice, when even one who thinks the action permissible is considered to have transgressed inadvertently. But since it is written "and she breaks faith with him," i.e., her husband, the deliberate willfulness and the unwittingness relate to her betrayal of her husband. This is similar to cases of murder, where inadvertence relates to the act of murder, and she who believes that it is permissible is considered virtually deliberate in betrayal of her husband. Colon did not make up his own exegesis, but explained and defined the type of inadvertent error that makes her permissible to her husband.[122]

120 *Tosafot*, Makot 7b s.v. *ela prat leomer mutar*.

121 Here the court is attempting to answer Ayash's critique of Colon, discussed above near note 82ff. Azoulay had a similar explanation, as is noted there, near notes 87 and 88.

122 RCJ, *op. cit.* (see note 113), p. 190.

The court thus distinguishes between different types of inadvertent error on the basis of the language used in the biblical verse in question. Just as unwittingness concerning the act itself is considered a mistake of fact, so with regard to the verse concerning the wayward woman which refers to the husband and not to the Almighty – the only error which is forgiven is factual error.

On these grounds the court rules that the two parties are forbidden to marry one another. Even if the argument was put forward, adds the court, that Colon's opinion concerns a woman who knows she is married and thinks it is permissible to commit adultery, while in this case the parties thought the woman was divorced, and thus unwittingly betrayed the husband, nevertheless, "in the final analysis the woman knew she must receive a writ of divorce (*get*) in order to be considered divorced."[123] However, this is a question of proof: if the woman could bring evidence to prove she was mistaken about this, she would be permitted according to Colon. The court therefore considered the woman's error to be one that hinged on her knowledge of the validity of her divorce – that is, it saw her error as a mistake of law.[124]

5. Inadvertent Adultery: Mistake of Law and Mistake of Fact

Although the rabbinical court did not use accepted legal terminology, it would appear that in its view, too, the deciding factor is whether the case involves mistake of fact or mistake of law. When the woman's error is one of fact (in the words of the rabbinical court, when the unwittingness is in the "act"), the woman is considered to have transgressed inadvertently and is not forbidden to her husband. If, however, the error is in law, that is, due to ignorance of the law (in the court's language, the error is in the "transgression"), she is not considered an inadvertent transgressor and is forbidden to her husband.

Why do the laws pertaining to inadvertent error with respect to adultery and murder differ from other laws of inadvertent error? We saw that a variety of attempts were made to explain the distinction by exegesis or grammatical analysis of the verses in which the term "inadvertent error" appears. However, these methods only resolve technical–legalistic questions concerning

123 *Ibid.*

124 At the end of its ruling (pp. 190–192), the court deals with the prohibition of her marrying her lover and its connection to the prohibition on her remaining with her husband, questions outside our purview here.

the formal linguistic grounds for the distinction. They do not explain why Jewish law established one set of rules for inadvertent error pertaining to the laws of the Sabbath and prohibited fat and another set for such error in the context of adultery and murder. To clarify this aspect of the subject, it will be helpful to examine contemporary legal theory on the distinction between mistake of law and mistake of fact in the criminal law.

The rationale for the principle that mistake of fact, if innocent and reasonable, relieves the transgressor of criminal liability, is evident. The transgressor, as it were, performed a prohibited action because he was convinced, innocently and reasonably, that it was legal and permissible. It can be assumed that had he been cognizant of the facts, he would not have perpetrated the deed. Thus from a moral point of view, this individual is not a criminal, and there is no reason for society to punish him by way of the criminal law.[125]

It may seem that the same claim can be made regarding mistake of law. The transgressor did not know the law; had he known, he would not have committed the criminal act. However, unlike mistake of fact, mistake of law is not deemed a valid defense, for the situation here is by no means parallel, as scholars have noted. Three basic rationales for the principle "ignorance of the law is no excuse" have been offered.

According to Austin, if ignorance of the law were a valid defense, every criminal defendant would claim ignorance, and the truth of these claims would be difficult to verify. This argument hinges on the issue of evidence, and does not get to the substance of the matter, for one could claim that were this defense to be accepted in courts of law, the rules concerning a plea of ignorance could be formulated in such a way that the burden of proof would fall on the defendant. Furthermore, it would be difficult for the defendant to prove lack of knowledge of the law and thus the success of this line of defense would be limited – many who did not know the law but were unable to prove their lack of knowledge would be judged guilty. We can thus discard this argument, and restate the question: why is ignorance of the law not an acceptable defense?[126]

---

125 See S.Z. Feller, *Elements of Criminal Law* (Hebrew), (Jerusalem: 1987), vol. 2, pp. 515ff.; idem, "'Mistake' in Israeli Criminal Law," *Reports to the Tenth International Congress of Comparative Law* (Jerusalem: 1978), pp. 207, 209ff.

126 For Austin's reasoning, and Holmes's critique of Austin, see O.W. Holmes, *The Common Law* (Boston: 1881), pp. 47–48. Also see G. Williams, *Criminal Law*, 2nd ed., (London: 1961), p. 290.

Holmes justifies the principle on didactic grounds. Accepting ignorance of the law as an adequate defense would reward the individual who did not learn the law. Only those who took the trouble to learn the law would be subject to punishment, while those who avoided such study would thereby evade criminal liability.[127]

The third argument, by Hall, is ethical-social in character. Criminal laws, Hall argues, reflect accepted societal norms. Accepting ignorance of the law or mistake in the law as a defense would be tantamount to allowing the criminal's personal norms to override societal norms. Society would no longer be capable of defending its values, or that defense would be woefully inadequate, since societal norms would govern only those familiar with them. Those who disregarded or sought to replace them would be able to avoid the penalties prescribed by law, for they would claim in their defense that they did not know the law.[128]

While these arguments are convincing, and indeed, in most legal systems it is an accepted principle of criminal law that claims of ignorance of, or mistake in, the law, are not admissible as arguments for the defense,[129] in Jewish law the reverse appears to be true. In most cases ignorance of the law is considered inadvertent error and dealt with like mistake of fact. Only in cases of

---

127 O.W. Holmes, *ibid.*, Lecture II, pp. 48ff.; G. Williams, *ibid.*, p. 289; cf. Feller, *op. cit.*, p. 550.

128 J. Hall, *General Principles of Criminal Law*, 2nd ed., (Indianapolis: 1947), pp. 382ff. Also see Feller, *ibid.*, pp. 549–550. On mistake of law and justification of the "ignorance of the law is no excuse" principle, see J.W. Salmond, *Jurisprudence*, 12th ed., (London: 1966), pp. 395–396.

129 It should be understood that there is no presumption that everyone is fully cognizant of the law, but only that ignorance of the law is not an acceptable defense. In Appeal of Legal Council 8/57, A Lawyer v. The Chair and Members of the Legal Council, P.D. 12, pp. 900, 909, Justice Silberg stated:

> According to accepted opinion, this maxim [i.e., *ignorance of the law is no excuse*] does not entail the 'irrefutable' presumption that everyone knows the law, or even just the criminal law. Its implication for one accused of a crime is solely that he cannot avail himself of the claim that, because he did not know the law, he is absolved of criminal responsibility for the crime he committed and he should go unpunished.

See also Feller, "Mistake ... ", *op. cit.*, pp. 219–220; cf. G. Williams, *op. cit.*, pp. 289–290.

adultery and murder is the situation different: here, ignorance of the law is not construed as inadvertent error.

In the context of the laws of the Sabbath, sacrifices or eating prohibited fat, transgressions committed by individuals are offenses against the Almighty. Let us consider the status of mistake of law and fact here. All agree that in the case of mistake of fact there is no point in punishing the individual involved, who did not intend to "wrong" the Almighty.

The same reasoning would seem, I believe, to apply to one who sins due to mistake of law. The sinner who does not know that he is transgressing does not "wrong" the Almighty, for he does not know that his act constitutes a sin against the Almighty. Under these circumstances, from the point of view of Jewish law, it is sufficient for him to bring a sin-offering, and, as in the case of other inadvertent transgressions, his sin is expiated.

The three rationales presented above to explain why mistake of law is not an admissible defense in criminal law do not apply in the context of laws between the individual and the Almighty. The argument from evidence fails, since even if in this world it is difficult to prove in court that someone knew the law, though he claims he did not, and is therefore acquitted, given divine omniscience, he will not escape judgment and punishment after his death.

The second justification, the didactic, is also inapplicable. Although much study is necessary if the law is to be mastered so that all the commandments and prohibitions can be observed,[130] one who remains unlearned is not punished for that alone. Tractate Shabbat gives as a classic example of ignorance of the law the case of "the infant who was abducted by non-Jews." There is no basis whatsoever for considering this individual, who violated many Sabbaths, a deliberate transgressor.

The third justification, though the strongest of the three, also seems inapplicable to laws governing relations between the individual and the Almighty. Society has an interest in mandating observance of its principles and norms, so that individuals will not repudiate them on the grounds that their own values are preferable to those of the community. This does not present a problem from the divine perspective, for the norms set down in Jewish law do not derive their authority from societal consensus. Even if they are transgressed daily, this authority remains intact. Societal norms, however, must be heeded, because their legitimacy rests on universal compliance. It is therefore clear

130 Compare: "The ignoramus is not fearful of sinning, and the boor is not pious" (Mishnah Avot 2:5).

why Jewish law accepts mistake of law as a defense in cases of transgression against divine injunctions such as laws of the Sabbath and prohibited fat, and views one who is ignorant of the law as an inadvertent sinner.

Adultery and murder differ fundamentally from violation of laws pertaining to the Sabbath and prohibited fat. Although adultery and murder are transgressions of divine injunction, their prohibition also has social ramifications. The person who violates the prohibitions against adultery and murder does not sin against Heaven alone, but also, unlike the desecrator of the Sabbath, whose act has repercussions only for the relationship between himself and the Almighty, affects other individuals or society as a whole. In cases of transgressions with societal implications, Jewish law espouses the reasoning behind the principle that ignorance of the law is no excuse, and does not classify mistake of law or ignorance of the law as inadvertent error.

When transgressions are between the individual and Heaven, the person who transgressed without knowing that his action was prohibited is an inadvertent transgressor. The law is content to exact an action signifying contrition, such as bringing an offering, and does not punish him as it does one who sins deliberately, since the transgressor had no criminal intent to violate biblical injunctions. In this sense such an individual is like one who makes a mistake of fact. Neither intends to transgress and both atone by bringing sacrifices.

But consider what happens when, on the other hand, in addition to the sin itself, a transgression has implications for a third party (say, the husband of the adulteress) or for society as a whole (as in cases of murder). No society can exist when its members do not know its laws and, out of ignorance, make their own rules as they see fit. Therefore, to protect society and third parties, Jewish law maintains that in these situations, one who does not know the law is not considered an inadvertent transgressor, but as virtually equivalent to one who transgresses deliberately.

In the context of adultery, the husband can demand a divorce from his wife even if the act of adultery was committed out of her ignorance of the prohibition against adultery, because the woman cannot expect her husband to defer to her ignorance rather than to the societal norms instituted for his benefit. This may explain Colon's inferences from the verse "and she breaks faith with him," and the rabbinical court's resort to this inference.

In any event, it certainly clarifies the responsa of Adret, Colon and the rabbinical court.

Adret's responsum hints at the three reasons presented above for the principle "ignorance of the law is no excuse." Adret first states: "she should have checked, and since she did not check she is forbidden to both."[131] The responsibility for checking and being careful is the obligation to familiarize oneself with the law through study, the second rationale noted above. The end of his statement, "If so, we will permit all women who commit adultery, saying: she did not think she was forbidden to do this," can be interpreted in two ways. He may have had in mind the first rationale, which turns on the issue of evidence, though it seems more likely his argument was in essence that of the third rationale: accepting the claim of such a woman is tantamount to undermining the prohibition against adultery, putting in its place the individual declaration of each woman to the effect that she believed it was permissible. The woman discussed in Adret's responsum, who believed she was not married, though she was, made a mistake of law,[132] and since adultery, as mentioned above, is a social crime as well as an interpersonal betrayal, her husband can demand a divorce.

This elucidates Colon's responsum as well. A married woman who believes that there is no prohibition against fornicating with another man, is not an inadvertent transgressor. The woman does not know the law, and as a result of the fact that her ignorance of this law affects her husband – unlike ignorance of other laws which concern matters solely between the individual and the Almighty – she is not considered merely to have transgressed by mistake, and her husband can divorce her.

In the case brought before the contemporary rabbinical court, a mistake of law[133] was also made: the woman was under the impression that the court

---

131 See above at notes 42–45.

132 The woman's error regarding the validity of the marriage ceremony is an error in law and not in fact. This very example is used by Williams to explain the distinction between mistake of law and mistake of fact:

> A mistake as to whether a marriage has been celebrated may be either a mistake of fact or a mistake of law. It is a mistake of fact if no ceremony has been performed; a ceremony is a fact of which a cinematograph picture could be taken. But the mistake is one of law if, though the ceremony has been performed, there is a misunderstanding of the rules of law governing the validity of the ceremony (G. Williams, *op. cit.*, p. 287).

The case before Adret is akin to the second example, and is therefore a case of mistake of law.

133 There is one sentence in the court ruling which could be taken as implying that the court viewed the mistake as a mistake of fact:

> And if we wish to differentiate this case from that of Colon – there, the woman says that she is permitted; she knew she had a husband but thought

ruling, and not the receipt of the writ of divorce, ended her marriage. In view of this, she gave herself permission to have sexual relations with another man during the period between the ruling and her receipt of the writ of divorce. Here too the woman in essence asks that the norm she has set up be preferred over societal norms. Her request must therefore be denied, and court ruled understandably that she is not an inadvertent transgressor, but an adulteress, and hence, prohibited to her husband and to her lover.

6. Conclusion

The Talmud discusses two cases where the laws of the adulterous woman (that is, that she is prohibited to both her husband and her lover, that the financial clauses of her marriage contract are abrogated, etc.) do not apply. The first is that of the woman who commits adultery under duress, and the second is that of the woman who has sexual intercourse with another man thinking him to be her husband.

In the post-talmudic literature there are many discussions of the woman who commits adultery believing that it is not prohibited, or whose adultery is the result of some other mistake of law, such as error regarding the validity of her marriage ceremony or the date of her divorce.

We found that in these cases of mistake of law, the woman is not considered a *shogeget*, or inadvertent transgressor, while in cases of mistake of fact, she is. We attempted to explain the difference between adultery and murder, on the one hand, where mistake of law does not excuse the transgressor, and other violations of Jewish law where inadvertent error is acknowledged with reference to both mistake of law and mistake of fact.

I argued that when the transgression is between the individual and the Almighty, the concept of *shgaga* encompasses both mistake of fact and mistake of law, for the Bible punishes only those transgressors who had

---

> it was permitted to commit adultery, so she is included under the category of "and she breaks faith with him," which is not the case here, for they erred and thought that although she was not actually divorced, the court ruling made her permissible to another and this is like inadvertent betrayal of the husband, for they believed erroneously that they were as good as divorced for this purpose (RCJ, p. 190).

However, as we explained in the preceding note, this is an error of law and not of fact.

“criminal intent” in their hearts. If the transgressor erred, whether in the law or regarding facts, he atones by bringing a sacrificial offering. But with regard to transgressions which go beyond the category of relationships between the individual and the Creator, and affect society as a whole (e.g., murder) or individuals within the society (e.g., the betrayed husband), mistake of law is not forgiven. In these areas, the transgressor who does not know the law, or errs concerning it, is not considered to have erred inadvertently. In this, Jewish law is like other legal systems.

Therefore, the woman who has sexual relations with another man, mistakenly believing him to be her husband, is permitted to her husband. On the other hand, the woman who has relations with another man thinking it permissible to commit adultery or erroneously thinking she is not married, or that she is divorced, errs in law. She is not considered an inadvertent transgressor, and, like the woman who deliberately commits adultery, is forbidden both to her husband and to her lover.

Contributors

Professor Hanina Ben-Menahem teaches at the Faculty of Law at The Hebrew University of Jerusalem, and is head of its Institute for Research in Jewish Law.

Moshe Drori is an attorney in Jerusalem.

Dr. Shimshon Ettinger teaches at the Faculty of Law at Bar Ilan University.

Professor Neil S. Hecht teaches at the Boston University School of Law, and is director of its Institute of Jewish Law.

Professor Berachyahu Lifshitz teaches at the Faculty of Law at The Hebrew University of Jerusalem.

Professor Shmuel Shilo teaches at the Faculty of Law at The Hebrew University of Jerusalem.

Professor Israel Ta-Shma teaches in the Talmud Department at The Hebrew University of Jerusalem.

Dr. Zerah Warhaftig is a former Minister of Religious Affairs for the State of Israel.

Contributors